CPC UCW
LLYFRGELL
LIBRARY
ABERYSTWY

AF606618

Progress in Mathematics
Vol. 33

Edited by
J. Coates and
S. Helgason

Birkhäuser
Boston · Basel · Stuttgart

Claude Zuily

Uniqueness and Non-Uniqueness in the Cauchy Problem

1983

Birkhäuser
Boston • Basel • Stuttgart

Author:
Claude Zuily
Mathématiques
Université de Paris-Sud
Centre d'Orsay
91405 ORSAY Cedex
France

Library of Congress Cataloging in Publication Data

Zuily, Claude, 1945-
Uniqueness and non-uniqueness in the Cauchy problem.

(Progress in mathematics ; v. 33)
Bibliography: p.
1. Cauchy problem — Numerical solutions.
2. Operator theory. I. Title II. Series:
Progress in mathematics (Cambridge, Mass.); 33.
QA377.Z83 1983 515.3'53 83-8743
ISBN 3-7643-3121-6

CIP-Kurztitelaufnahme der Deutschen Bibliothek

Zuily, Claude:
Uniqueness and non-uniqueness in the Cauchy problem / Claude Zuily. - Boston ; Basel ; Stuttgart : Birkhäuser, 1983.
(Progress in mathematics ; Vol. 33)
ISBN 3-7643-3121-6

NE: GT

ISBN 3-7643-3121-6
Printed in USA

To Yasmine, Estelle and Stephane

INTRODUCTION

The question of uniqueness of the non-characteristic Cauchy problem, in the C^{∞} framework, for more general operators than the hyperbolic ones, goes back to T. Carleman [11] who considered, in the two dimensional case, operators with real coefficients and simple complex characteristics. His proof, which was based on weighted estimates, turned out to be quite powerful since this method (called "Carleman estimates method") with various modifications is still today the only tool used to prove uniqueness theorems.

The first general results of the theory were given in 1957 by A. P. Calderon [8] as an interesting application of singular integral operators (or pseudo-differential operators). After that paper, several works were done (L. Hörmander [17], [18], S. Mizohata [32], R. Pederson[38], M. Protter [46]) giving new sufficient conditions for uniqueness and mainly devoted to elliptic operators.

Some time later, important progress was made by P. Cohen [12] and A. Pliš [40] to [45]. They gave several counterexamples which showed in particular that uniqueness does not hold for arbitrary elliptic operators, but depends of the multiplicities of the complex characteristics ; it depends also on all terms of the operator and on regularity of the coefficients. These authors gave, moreover, a way to construct non-trivial solutions, but this method was applied to particular examples or non invariant situations.

Then, L. Hörmander in his book, [19] chap. VIII, introduced the very important notion of strongly pseudo-convex hypersurfaces and proved uniqueness results for principally normal and elliptic operators.

In 1973 K. Watanabe [55] extended Calderon's theorem to certain operators with triple characteristics (the general case appeared later, in [57]).

L. Hörmander [20], trying to unify the counterexamples discussed above, gave in 1975 a more systematic way to construct non trivial solutions based on a very delicate variant of the geometrical optics method. However the results given there concerned perturbations of constant coefficient operators.

Since this time, using that method and the "Carleman estimates" it has been possible to go further. Indeed, the works of S. Alinhac-C. Zuily [5], R. Lascar-C. Zuily [27], S. Alinhac [1], [2], L. Hörmander [19] (Chap VIII) give almost necessary and sufficient conditions for uniqueness based on geometrical properties. Let us also mention the geometrical non-uniqueness theorem of S. Alinhac-M.S. Baouendi [4].

The aim of these notes is to present some aspects of the theory and to describe the tools used, mainly the non-uniqueness constructions. The plan is the following.

In the first Chapter, first order differential operators in $\mathbb{R}^n$ are considered. Using the works of M. Strauss-F. Treves [51] (with a simple proof) and S. Alinhac [1], an almost necessary and sufficient condition for uniqueness (for every zero order perturbation) is proved when $n \geqslant 3$. The two dimensional case is also discussed with positive results and counterexamples.

Chapter Two is devoted to the works initiated by Calderon's theorem. It begins by an improvement of his original work in the case of smooth characteristics. A case of non-smooth roots is also discussed (C. Zuily [65]). Then the result of K. Watanabe is given and a counterxample of A. Plǐs [4] is discussed in detail. This chapter ends with some results concerning operators with characteristics of arbitrary high multiplicity. (K. Watanabe-C. Zuily [59], C. Zuily [64] etc...).

The third chapter begins by the operators with real principal part and Hörmander's uniqueness theorem ([19]) for strongly pseudo-convex hypersurfaces is proved. However the proof given here, due to N. Lerner [29], is a little more transparent, thanks to the use of the Weyl calculus for (pseudo) differential operators and an appropriate choice of the weight functions in the Carleman estimate. A non-uniqueness result (Alinhac [1]) is proved when the strong pseudo-convexity is violated (in a strong sense). Then we deal with a class of quasi-homogeneous second order operators and a geometrical (almost) characterization is given in terms of an appropriate notion of pseudo-convexity (R. Lascar-C. Zuily [27]). The third section of this chapter is devoted to the simplest case of operators with double real characteristics and a quite complete (and surprising) answer is given. (S. Alinhac-C. Zuily [5]). In section four we deal with elliptic operators; Hörmander's uniqueness theorem is proved by the method of N. Lerner [29] and a non-uniqueness result of S. Alinhac [2] is then stated. We end this chapter by several general non-uniqueness results stated without proof. ([1],[4],[20]).

In these notes we have considered only classical solutions, but several results could be extended to arbitrary distribution solutions (see for instance F. Cardoso - J. Hounie [10]). On the other hand we have not tried to give the minimal smoothness assumptions on the coefficients and these could be weakened. Moreover some material is missing : for instance uniqueness for constant coefficient operators (P.M. Goorjan [16], K. Watanabe [55] etc...) or non-uniqueness results in the case of non-smooth coefficients (in

Hölder classes for instance). For these kind of results we refer to P. Cohen [12], A. Pliš [44], L. Hörmander [20]. Let us remark that the material given here can be used in the characteristic Cauchy problem.

The bibliography at the end has been made as complete as possible, as far as the title of this book in its (more or less) strict sense is concerned.

A preliminary version of a major part of this book was written as lecture notes at University of Pernambuco at Recife (Brazil) during the summer 1981, while the author was visiting that University, supported by C.N.P.Q. and F.I.N.E.P.

I would like to thank Prof. F. Cardoso and his collegues in the Mathematical department for their hospitality and the p.d.e staff of this department for making this visit so stimulating.

My thanks also to N. Lerner for useful discussions concerning Section 1 and 4 of chapter 3, to Mrs. Neida Maria Santos de Souza for typing early draft material and to Mrs. Simonot who prepared the final camera-ready copy.

Orsay, October 1982.

CONTENTS

Chapter One

FIRST ORDER DIFFERENTIAL OPERATORS

We shall deal here with a first order differential operators, in an open set Ω of $\mathbb{R}^n$, with complex valued coefficients :

(0.1) $$L = P_1 + iP_2 + c$$

where for $j=1,2, P_j = \sum_{k=1}^{n} a_{jk}(x)\, D_k$ has real symbol $p_j(x,\xi)$. Here D_j, as usual, stands for $\frac{1}{i}\frac{\partial}{\partial x_j}$.

Let S be a hypersurface in Ω, given in a neighborhood V of a point $x_o \in \Omega$ by

(0.2) $$S = \{x \in V : \phi(x) = \phi(x_o)\}$$

with a C^1 function ϕ such that $d\phi(x_o) \neq 0$.

The hypersurface will be assumed to be non-characteristic for P at x_o, which means here that

(0.3) $$P_1\phi(x_o) + iP_2\phi(x_o) \neq 0.$$

It will be convenient, in that follows, to work in coordinates where L and S take simple forms. Indeed, we have the following easy lemma, whose proof is left to the reader.

Lemma 0.1 : There exist coordinates $(y_1,\dots,y_{n-1},s)$ near the origin of $\mathbb{R}^n$, where $y = (y_1,\dots,y_{n-1}) \in \mathbb{R}^{n-1}$ and $s \in \mathbb{R}$, in which S and L take the form

(0.4) $$\begin{cases} S = \{(y,s) \in V : s=0\} \\ L = a(y,s)\,\{D_s + i\sum_{j=1}^{n-1} b_j(y,s)\, D_{y_j} + c\} \end{cases}$$

with $a(0,0) \neq 0$ and the b_j are real.

We shall start the study of uniqueness by a very simple case, namely,

§1. THE ELLIPTIC CASE

It means that we have

(1.1) $p_1(x,\xi) + ip_2(x,\xi) \neq 0 \quad \forall(x,\xi) \in V \times \mathbb{R}^n \setminus 0.$

This implies obviously that $n \leq 2$. The case $n=1$ being trivial, we shall suppose $n=2$.

<u>Theorem 1.1 : Let</u> $L^o = P_1 + iP_2$ <u>be elliptic. Suppose that the coefficients of</u> $P_j (j=1,2)$ <u>are Lipschitz continuous in</u> V <u>and</u> S <u>is a</u> C^1 <u>hypersurface. Then there exists a neighborhood</u> W <u>of</u> x_o <u>such that if</u> u <u>is a</u> C^1 <u>function in</u> V <u>satisfying, with</u> $C = C(u) \geq 0$,

(1.2) $$\begin{cases} |L^o u(x)| \leq C|u(x)| \\ u=0 \quad \text{in} \quad \{\phi(x) < \phi(x_o)\} \end{cases}$$

<u>then</u> u <u>vanishes in</u> W.

Let us remark that this theorem implies obviously the uniqueness of the Cauchy problem for $L = P_1 + iP_2 + c$ if $c \in L^\infty(V)$. However it can also be applied to non-linear problems $Lu = P_1u + iP_2u + f(u) = 0$.

<u>Proof of theorem 1.1</u> : According to lemma 0.1 we will work in the (y,s) coordinates in which

$$L^o = \alpha(y,s) \{\frac{\partial}{\partial s} + i\beta(y,s) \frac{\partial}{\partial y}\}$$

with $\alpha(0,0) \neq 0$. The ellipticity of L^o means that $\beta(0,0) \neq 0$. The change of coordinates $x=y$, $t=s+x^2$ make the support of the solution $\tilde{u}$ contained in the strictly convex set $\{(x,t) : t > x^2\}$; the operator L^o becomes

(1.3) $$\tilde{L}^o = f(x,t) \{\frac{\partial}{\partial t} + (a+ib)(x,t) \cdot \frac{\partial}{\partial x}\}$$

with $f(0,0) \neq 0$ and $b(0,0) \neq 0$.

The theorem 1.1. will follow from the following Carleman estimate.

Proposition 1.2 : *There exist positive constants* C, r *such that for* k *and* T^{-1} *big enough we have*

$$(1.4)\qquad \int_0^T e^{k(t-T)^2} ||v||^2 dt \leq \frac{C}{k} \int_0^T e^{k(t-T)^2} ||\tilde{L}^o v||^2 dt$$

for every $v \in C^1$ *such that* $\operatorname{supp} v \subset \{(x,t) : 0 \leq t \leq T, |x| \leq r\}$. *Here* $|| \ ||$ *denotes the* $L^2(\mathbb{R}^{n-1})$ *norm.*

We first show how to deduce the uniqueness result from (1.4). Let $\tilde{u} \in C^1$ be a solution of $|\tilde{L}^o \tilde{u}(x,t)| \leq C |\tilde{u}(x,t)|$ with support in $\{(x,t) : t \geq x^2\}$. Let $\theta(t)$ be a C^∞ real function such that $\theta(t) = 0$ for $t > T$, $\theta(t) = 1$ for $t \leq T_1 < T$ and $0 \leq \theta \leq 1$. Take $0 < T_2 < T_1$ and apply (1.4) to $v = \theta\tilde{u}$. We get

$$\int_0^T e^{k(t-T)^2} ||\theta\tilde{u}||^2 dt \leq \frac{c}{k} \int_0^T e^{k(t-T)^2} ||\tilde{L}^o \theta\tilde{u}||^2 dt \leq$$

$$\leq \frac{c}{k} \int_0^T e^{k(t-T)^2} ||\theta\tilde{L}^o\tilde{u}||^2 dt + \frac{c}{k} \int_0^T e^{k(t-T)^2} ||[\tilde{L}^o, \theta]\tilde{u}||^2 dt.$$

Using the above inequality between $|\tilde{L}^o\tilde{u}|$ and $|\tilde{u}|$ and taking k big enough, we get

$$\int_0^T e^{k(t-T)^2} ||\theta\tilde{u}||^2 dt \leq \frac{c'}{k} \int_0^T e^{k(t-T)^2} ||[\tilde{L}^o, \theta]\tilde{u}||^2 dt.$$

which implies, by the choice of θ,

$$\int_0^{T_2} e^{k(t-T)^2} ||\tilde{u}||^2 dt \leq \frac{c'}{k} \int_{T_1}^T e^{k(t-T)^2} ||[\tilde{L}^o, \theta]\tilde{u}||^2 dt,$$

from which we get, since $e^{k(t-T)^2}$ is decreasing,

$$\int_0^{T_2} ||\tilde{u}||^2 dt \leq \frac{c'}{k} e^{k\{(T_1-T)^2-(T_2-T)^2\}} \int_{T_1}^T ||[\tilde{L}^o, \theta]\tilde{u}||^2 dt.$$

Letting k go to $+\infty$ we deduce that $\tilde{u}=0$ for $0 \leq t \leq T_2$.

Proof of Proposition 1.2 : It is of course equivalent to deal with $\frac{1}{f}\tilde{L}^o = \frac{\partial}{\partial t} + (a+ib)\frac{\partial}{\partial x}$. Putting $e^{\frac{k}{2}(t-T)^2} \cdot v = w$

we get

$$e^{k(t-T)^2} \frac{1}{f} \tilde{L}^o v = \left(\frac{\partial}{\partial t} - k(t-T) + (a+ib)\frac{\partial}{\partial x}\right) w = Aw + Bw,$$

where

$$(1.5) \qquad \begin{cases} A = \dfrac{\partial}{\partial t} + a\dfrac{\partial}{\partial x} \\ B = -k(t-T) + ib\dfrac{\partial}{\partial x} \end{cases} .$$

We then have

$$(1.6) \qquad \int_o^T e^{k(t-T)^2} \left\|\frac{1}{f} \tilde{L}^o v\right\|^2 dt = \\ = \int_o^T \|Aw\|^2 dt + \int_o^T \|Bw\|^2 dt + 2\mathrm{Re}\int_o^T (Aw,Bw)\,dt .$$

Now it is easy to see, integrating by parts, that

$$(1.7) \qquad 2\mathrm{Re}\int_o^T \left(\frac{\partial w}{\partial t}, -k(t-T)w\right) dt = k\int_o^T \|w\|^2 dt$$

$$(1.8) \qquad \left| 2\mathrm{Re}\int_o^T \left(a\frac{\partial w}{\partial x}, -k(t-T)w\right) dt \right| \leqslant ckT\int_o^T \|w\|^2 dt$$

$$(1.9) \qquad 2\mathrm{Re}\int_o^T \left(a\frac{\partial w}{\partial x}, ib\frac{\partial w}{\partial x}\right) dt = 0 .$$

On the other hand ,

$$\int_o^T \left(\frac{\partial w}{\partial t}, ib\frac{\partial w}{\partial x}\right) dt = -\int_o^T \left(w, ib'_t\frac{\partial w}{\partial x}\right) dt + \int_o^T \left(w, ib'_x\frac{\partial w}{\partial t}\right) dt - \int_o^T \left(ib\frac{\partial w}{\partial x}, \frac{\partial w}{\partial t}\right) dt;$$

from which we deduce since b does not vanish ,

$$2\mathrm{Re}\int_o^T \left(\frac{\partial w}{\partial t}, ib\frac{\partial w}{\partial x}\right) dt = -\int_o^T \left(\frac{b'_t}{b} w, ib\frac{\partial w}{\partial x}\right) dt + \int_o^T \left(b'_x w, i\frac{\partial w}{\partial t}\right) dt = 1/+2/.$$

From (1.5) we get

$$1/ = -\int_o^T \left(\frac{b'_t}{b} w, Bw\right) dt - k\int_o^T (t-T)\left(\frac{b'_t}{b} w, w\right) dt$$

$$2/ = \int_0^T (b_x' w, iAw)\,dt - \int_0^T (\frac{b_x' a}{b} w, ib \frac{\partial w}{\partial x})\,dt$$

$$2/ = \int_0^T (b_x' w, iAw)\,dt - \int_0^T (\frac{b_x' a}{b} w, Bw)\,dt - k \int_0^T (t-T)(\frac{b_x' a}{b} w, w)\,dt.$$

Using the Cauchy Schwartz inequality and $a.b \leqslant \varepsilon a^2 + \frac{1}{4\varepsilon} b^2$, we get

$$(1.10) \qquad \left| 2\mathrm{Re} \int_0^T (\frac{\partial w}{\partial t}, ib \frac{\partial w}{\partial x})\,dt \right| \leqslant$$

$$\leqslant \varepsilon \int_0^T (\|Aw\|^2 + \|Bw\|^2)\,dt + c_\varepsilon \int_0^T \|w\|^2 dt + ckT \int_0^T \|w\|^2\,dt.$$

From (1.6) to (1.10) we get

$$\int_0^T \|Aw\|^2\,dt + \int_0^T \|Bw\|^2\,dt + k \int_0^T \|w\|^2\,dt \leqslant$$

$$\leqslant \int_0^T e^{k(t-T)^2} \|\frac{1}{f} \tilde{L}^o v\|^2 dt + \varepsilon \int_0^T (\|Aw\|^2 + \|Bw\|^2)\,dt +$$

$$+ c_\varepsilon \int_0^T \|w\| \, dt + ckT \int_0^T \|w\|^2\,dt.$$

Choosing $\varepsilon = \frac{1}{2}$ and taking k and T^{-1} big enough, we deduce (1.4) from the above inequality. This ends the proof of Theorem 1.1.∎

We are now going to extend the uniqueness result to more general first order operators, namely those which satisfy the condition (P) of L.Nirenberg - F. Treves which occurs in the local sovability theory.

Before starting, let us recall this condition and discuss its meaning.

Let $L = P_1 + iP_2 + c$ as before and suppose by (0.3) that the coefficients of P_1 do not vanish together at x_o (which is always satisfied by L or iL). We state :

$$(P)\begin{cases}\text{In a neighborhood } V \text{ of } x_o, \text{ the vector field } \frac{1}{i}P_2 \\ \text{does not change direction along any characteristic} \\ \text{curve of } \frac{1}{i}P_1 \text{ contained in } V.\end{cases}$$

It is easy to see that in the (y,s) coordinates selected by Lemma 0.1, where

$L = \frac{1}{i}\, a(y,s)\, \{\frac{\partial}{\partial s} + \sum_{j=1}^{n-1} b_j(y,s)\, \frac{\partial}{\partial y_j} + c\}$, the condition (P) reads

$$(\tilde{P})\begin{cases}\text{There is an open neighborhood } \omega \text{ of the origin in } \mathbb{R}^{n-1}, \\ \text{a number } T > 0 \text{ and a mapping } y \mapsto v(y) \text{ from } \omega \text{ into the} \\ \text{unit sphere } S^{n-2} \text{ such that} \\ \\ b_j(y,s) = (\sum_{j=1}^{n-1} b_j^2(y,s))^{1/2}.\ v_j(y) \qquad \forall (y,s) \in \omega \times]-T,T[\ .\end{cases}$$

We will note $\vec{b}(y,s) = (b_j(y,s))_{j=1...n-1}$ and

$$|\vec{b}(y,s)| = (\sum_{j=1}^{n-1} b_j^2)^{1/2}.$$

Now when $p_1(x_o,\xi)$ and $p_2(x_o,\xi)$ are linearly independant it turns out that condition (P) is equivalent to

(1.11) There exists f,g in $C^\infty(V)$ such that $\{p_1,p_2\} = fp_1 + gp_2$.

The statement being invariant, let us check this in the (y,s) coordinates. Since $\sum_{j=1}^{n-1} |b_j(0,0)| \neq 0$ it follows that $|\vec{b}| \in C^\infty$.

If $(\tilde{P})$ is satisfied then

$$\frac{\partial b_j}{\partial s} = \frac{\partial |\vec{b}|}{\partial s}\, v_j(y).$$

It follows that

$$\left[\frac{\partial}{\partial s}, \sum_{j=1}^{n-1} b_j(y,s)\frac{\partial}{\partial y_j}\right] = \frac{\frac{\partial|\vec{b}|}{\partial s}}{|\vec{b}|} \sum_j b_j(y,s)\frac{\partial}{\partial y_j} .$$

On the other hand, if the above equality is true, the function b_j has to satisfy $\frac{\partial b_j}{\partial s} = \alpha\, b_j$, which implies that condition $(\tilde{P})$ is satisfied.

Let us note then, that condition (P) means that :

(1.12) $\{p_1, p_2\} = 0$ on $\Sigma = \{(x,\xi) \in V \times \mathbb{R}^n \setminus 0 : p_1(x,\xi) = p_2(x,\xi) = 0\}$

where { , } denotes the Poisson bracket.
We can now discuss the question of uniqueness.

§2. THE UNIQUENESS THEOREM UNDER CONDITION (P)

The main result of this section is the following :

Theorem 2.1 : *Let us suppose that* $L = P_1 + iP_2 + c$ *has* C^1 *coefficients and satisfies condition* (P) *in a neighborhood* V *of a point* x_0 *belonging to the* C^1 *hypersurface* S. *There exists a neighborhood* W *of* x_0 *such that if* $u \in C^1(V)$ *satisfies*

$$\begin{cases} Lu = 0 \text{ in } V \\ u|_S = 0 \end{cases}$$

then $u \equiv 0$ *in* W .

Proof : We are going to work in the (y,s) coordinates, selected by Lemma 0.1, in a neighborhood $\omega_0 \times]-r,r[\times]-T,T[$ of the origin. Theorem 2.1 will be a consequence of the following two dimensional case, namely:

Theorem 2.2 : *Consider in a neighborhood* $Q =]-r,r[\times]-T,T[$ *of the origin in* $\mathbb{R}^2$ *the operator*

$$L_1 = \frac{\partial}{\partial t} + ib(x,t)\frac{\partial}{\partial x} + c(x,t)$$

and suppose

i) $b(x,t) \geqslant 0 \ \forall (x,t) \in Q$

ii) $D_x^\alpha b \in L^\infty(Q)$, $|\alpha| \leqslant 1$, $D_x^\alpha c \in L^\infty(Q)$, $|\alpha| \leqslant 1$.

Let u *be a* C^1 *function in* Q *such that* $L_1 u = 0$ *in* Q *and* $u_{|t<0} = 0$. *Then* u *vanishes in* Q.

Let us first show how to deduce Theorem 2.1 from Theorem 2.2. Recall that L is given by (0.4) and satisfies condition $(\tilde{P})$. Consider the set

$$N_o = \{y \in \omega_o : \forall s : |s|<T, \ \vec{b}(y,s) = 0\}.$$

If $y_o \in \omega_o \setminus N_o$, let us make a change of coordinates in the y-variables near y_o in order to transform $\sum_{j=1}^{n-1} v_j(y) \frac{\partial}{\partial y_j}$ into $\frac{\partial}{\partial x_1}$. The operator L becomes, by $(\tilde{P})$, up to a non vanishing factor :

$\frac{\partial}{\partial s} + i|\vec{b}| \frac{\partial}{\partial x_1} + c$ where $(x_2,\dots,x_n)$ play a role of parameters. By theorem 2.2 u vanishes in $\omega_o \setminus N_o \times]-T,T[$. Now in $\{y_o\} \times]-T,T[$, with $y_o \in N_o$, the equation $Lu = 0$ becomes $\frac{\partial u}{\partial s}(y_o,s) = 0$; since $u|_{s=0} = 0$ it follows that $u(y_o,s) = 0$ for $s \in]-T,T[$.

Proof of theorem 2.2 : We prove first a Carleman estimate. Let $f_o(x)$, $f_1(t)$, $\theta(x)$ be such that $D_x^\alpha f_o \in L^\infty(]-r,r[)$, $|\alpha| \leqslant 2$, $D_x^\alpha \theta \in L^\infty(]-r,r[)$, $|\alpha| \leqslant 2$ and $D_t^\alpha f_1 \in L^\infty(]-T,T[)$ for $|\alpha| \leqslant 1$. Consider the function

$$(2.1) \qquad f(x,t) = f_o(x) + \int_{\theta(x)}^t b(x,s) f_1(s)\, ds.$$

It follows that $D_x^\alpha f \in L^\infty(Q)$ for $|\alpha| \leqslant 2$ where $Q =]-r,r[\times]-T,T[$. We have the following :

<u>Proposition 2.3</u> : <u>Let us suppose that there exists</u> $C_o > 0$ <u>such that</u>

$$\frac{\partial f_1}{\partial t} + \frac{\partial}{\partial x}\left(b\,\frac{\partial f}{\partial x}\right) \geq C_o \quad \text{in} \quad Q. \tag{2.2}$$

<u>Then there exist positive constants</u> C, k_o, <u>such that for</u> $k \geq k_o$

$$k \iint_Q e^{2kf}|v|^2\,dxdt \leq C \iint_Q e^{2kf}\{|L_1 v.\partial_x \bar{v}| + k|L_1 v.\bar{v}|\} \tag{2.3}$$

<u>for every</u> $v \in C_o^1(Q)$.

<u>Proof</u> : Consider $c_1(x,t) = \int_o^t c(x,s)\,ds$ and $v = v_1\, e^{-c_1}$.

Then

$$L_1 v = \left[\frac{\partial}{\partial t} + ib\left(\frac{\partial}{\partial x} + d\right)\right] v_1 \;,\quad d = -\frac{\partial c_1}{\partial x} = h + ig\,.$$

Now, it is easy to see that if $v_1 = e^{-kf}w$, then we have

$$e^{kf}\left\{\frac{\partial v_1}{\partial t} + ib\left(\frac{\partial v_1}{\partial x} + dv_1\right)\right\} = Mw + ibNw$$

with

$$\begin{cases} M = \dfrac{\partial}{\partial t} - ibkf'_x + ibh \\[2mm] N = \dfrac{\partial}{\partial x} + ikf_1 + ig. \end{cases} \tag{2.4}$$

We see from (2.4) that

$$M^\star = -M, N^\star = -N \quad \text{and} \quad \frac{1}{i}\,[M,N] = k\left\{\frac{\partial f_1}{\partial t} + \frac{\partial}{\partial x}\left(b\frac{\partial f}{\partial x}\right)\right\} + \alpha,\ \ \alpha \in L^\infty(Q).$$

Now

$$-2\mathrm{Re}(Mw + ibNw,\ iNw) = 2\ \mathrm{Im}(Mw,Nw) + \iint_Q b|Nw|^2 dxdt.$$

On the other hand ,

$$(Mw,Nw) = -\,(w,\ [M,N]\,w) + (Nw,Mw)$$

which implies that $2i\ \mathrm{Im}(Mw,Nw) = -(w,\ [M,N]\,w)$.

It follows that

$$-2\mathrm{Re}(Mw + ibNw,\ iNw) = (w, \frac{1}{i}[M,N]w) + \iint_Q b\,|Nw|^2 dxdt.$$

Using (2.2) and the expression for $\frac{1}{i}[M,N]$ we deduce if k is big enough,

$$-2\mathrm{Re}(Mw + ibNw,\ iNw) \geqslant C\,k \iint_Q |w|^2\, dxdt,$$

which implies (2.3)∎

We are now able to prove theorem 2.2. We proceed by contradiction, assuming that if u is the function under study then $F = \mathrm{supp}\, u \neq \phi$.
There exists an interval $J \subset \mathbb{R}_x$ such that

(2.5) $\qquad \forall\, x \in J \quad \exists t < T : u(x,t) \neq 0.$

We choose now the function θ which appears in (2.1). We take $\theta \in C^2(J)$ such that

a) $(\mathrm{Graph}\ \theta) \cap F \neq \phi$

b) For every (x,t) in F $\quad t \geqslant \theta(x)$.

Such a θ can be easily constructed using parabolas.
Let $(x_o, \theta(x_o)) \in (\mathrm{Graph}\ \theta) \cap F$.

<u>Case 1</u>: $b(x_o, \theta(x_o)) \neq 0$.

It follows that L is elliptic near this point. The curve $t = \theta(x)$ is C^2 and u vanishes for $t < \theta(x)$. Applying theorem 1.1 we deduce that u must vanish in a neighborhood of $(x_o, \theta(x_o))$, which is a contradiction.

<u>Case 2</u> : $b(x_o, \theta(x_o)) = 0$.

Let us define

$$\bar{t}(x_o) = \sup\{t \geqslant \theta(x_o) : b(x_o, s) = 0,\ \theta(x_o) \leqslant s \leqslant t\}.$$

Then $\bar{t}(x_o) < T$. Indeed, suppose $\bar{t}(x_o) = T$;on the segment $\ell_o = [(x_o, \theta(x)), (x_o, \bar{t}(x_o))]$ b vanishes, giving

$\frac{\partial u}{\partial t}(x_o,t) = 0$, so $u(x_o,t) = 0$ for $t < T$ since $u(x_o,t)=0$ for $t < \theta(x_o)$. This cannot happen by (2.5).

We are going to use Proposition 2.3 in $J \times]-T,T[$ with

$$f_o(x) = -(x-x_o)^2 \ , \ f_1(t) = -e^{-t}$$

which gives

$$f(x,t) = -(x-x_o)^2 - \int_{\theta(x)}^{t} b(x,s)\ e^{-s}\, ds .$$

Let U_1 an open neighborhood of ℓ_o such that

$$\sup_{(x,t)\in U_1} \left|\frac{\partial}{\partial x}\left(b \frac{\partial f}{\partial x}\right)\right| \leq \frac{1}{2} e^{-T} . \tag{2.6}$$

Such a neighborhood exists because b vanishes on ℓ_o and, being non negative, $\frac{\partial b}{\partial x}$ vanishes at the points where b vanishes.

In U_1, since $t \leq T$, we have by (2.6)

$$\frac{\partial f_1}{\partial t} + \frac{\partial}{\partial x}\left(b \frac{\partial f}{\partial x}\right) = e^{-t} + \frac{\partial}{\partial x}\left(b \frac{\partial f}{\partial x}\right) \geq \frac{1}{2} e^{-T} .$$

It follows that the condition (2.2) is satisfied in U_1. Let U_2 be an open neighborhood of ℓ_o contained in U_1. Take $g \in \mathcal{D}(U_1)$ with $g \equiv 1$ on U_2. Consider

$$U_+ = \{(x,t) \in U_1 \ : \ f(x,t) \geq 0 \} .$$

We have

$$U_+ \cap F = \{x \in J,\ t \geq \theta(x) \ : \ (x-x_o)^2 + \int_{\theta(x)}^{t} b(x,s) e^{-s}\, ds \leq 0\}$$

and consequently

$$U_+ \cap F = \{x=x_o,\ \theta(x_o) \leq t \leq \bar{t}(x_o)\} = \ell_o .$$

Since $g = 1$ on U_2, the set $(\text{supp grad }(g)) \cap F$ does not intersect $U_+ \cap F$, so we have : $(\text{supp grad}(g)) \cap U_+ \cap F = \phi$.

Let us define

$$\begin{cases} V_\varepsilon^- = \{(x,t) \in U_1 \ : \ f(x,t) < -\varepsilon\} \\ V_\varepsilon^+ = \{(x,t) \in U_1 \ : \ f(x,t) \geq -\varepsilon\} . \end{cases} \tag{2.7}$$

It follows that

$$(\text{supp grad}(g)) \cap F \subset \bigcup_{\varepsilon > 0} V_\varepsilon^- .$$

By a compactness argument there exists $\varepsilon_o > 0$ such that

(2.8) $$(\text{supp grad}(g)) \cap F \subset V_{\varepsilon_o}^- .$$

Take $\varepsilon_1 \in]0, \varepsilon_o[$ and let U_3 be a neighborhood of ℓ_o such that

(2.9) $$U_3 \subset V_{\varepsilon_1}^+ \cap U_2 .$$

Such a neighborhood exists because

$$- f(x,t) = (x-x_o)^2 + \int_{\theta(x)}^t b(x,s) e^{-s}\, ds < \varepsilon_1$$

and b vanishes on ℓ_o.

Consider the function u under study and apply the inequality (2.3) of Proposition 2.3 to $v = g.u$. Taking $k \geqslant 1$ we get

$$\iint_{U_3} e^{2kf} |u|^2\, dxdt \leqslant \iint_{U_1} e^{2kf} \{ |Lgu . \partial_x(g\bar{u})| + |Lu . g\bar{u}| \} dxdt .$$

Now $Lgu = gLu + [L,g]u = [L,g]u$, from which it follows using (2.8) that

$$\text{supp}([L,g]u) \subset V_{\varepsilon_o}^- .$$

Using (2.7) and (2.9) we can write

$$\iint_{U_3} |u|^2\, dxdt \leqslant C(u) . e^{-2k(\varepsilon_o - \varepsilon_1)} .$$

Letting $k \to +\infty$ it follows that $u \equiv 0$ in U_3, which is a contradiction∎

Before we state the non-uniqueness result let us recall that according to (1.12), condition (P) means, when $p_1(x,\xi)$ and $p_2(x,\xi)$ are linearly independant, that

$$\{p_1, p_2\}(x,\xi) = 0 \quad \text{on } \Sigma = \{(x,\xi) \in V \times \mathbb{R}^n \setminus 0 : p_1(x,\xi) = p_2(x,\xi) = 0\}.$$

§3. THE NON-UNIQUENESS THEOREM

The main result of this section is the following.

Theorem 3.1 : Let $L = P_1 + iP_2 + c$ be a first order differential operator with C^∞ coefficients in a neighborhood of $x_o \in V$, such that $p_1(x,\xi)$ and $p_2(x,\xi)$ are linearly independant in $V \times \mathbb{R}^n \setminus 0$. Suppose that

(*) There exists $\xi^o \in \mathbb{R}^n \setminus 0$ such that $p_1(x_o,\xi^o) = p_2(x_o,\xi^o) = 0$ but $\{p_1,p_2\}(x_o,\xi^o) \neq 0$.

Then there exist $C^\infty(V)$-functions u and a with support in $\{x \in V : \phi(x) \geqslant \phi(x_o)\}$ and

$$(3.1) \qquad \begin{cases} Lu - au = 0 \quad \text{in } V \\ \\ x_o \in \operatorname{supp} u . \end{cases}$$

Before we give the details of the (quite long) proof of the theorem 3.1, let us explain its structure in giving the different steps of the construction of the functions u and a. Let us work in the (x,t) coordinates where L takes the form $L = \frac{\partial}{\partial t} + i \sum_{j=1}^{n-1} b_j \frac{\partial}{\partial x_j} + c$. We are going first to construct u, from which we deduce $a = \frac{Lu}{u}$. We shall take $u = \sum_{k \geqslant k_o} u_k$, each u_k having support contained in $\omega_x \times [b_{k+1}, b_{k-1}]$, (b_k) being a sequence of real numbers converging to zero when k goes to infinity. Following the geometrical optic method, u_k will be taken in $\omega \times [b_{k+1}, b_{k-1}]$ as

$$u_k(x,t) = e^{i\tau_k \xi(x,b_k)} \, e^{\nu_k \phi(x,s,b_k)} \, e^{-\gamma_k(x)} \, w(x,s,b_k) \;, \quad s = \frac{t-b_k}{b_k^\theta}$$

where τ_k, ν_k, θ are real parameters.

We begin by finding the phase functions ϕ and ξ satisfying convexity properties and good estimates with respect to k. After that, we fix the parameters and chose γ in order to localize the set $|u_k| = |u_{k+1}|$. Indeed, since in (b_{k+1}, b_k) one has $a = \frac{Lu_k + Lu_{k+1}}{u_k + u_{k+1}}$, we have to look at this set if we want the function a to be well defined. The function w is then found by solving transport equations. The u_k's are now such that $|u_k| \gg |u_{k+1}|$ near b_k, $|u_{k+1}| \gg |u_k|$ near b_{k+1} and $|u_k| = |u_{k+1}|$ on curves close to the mid-point of $[b_{k+1}, b_k]$. We then modify slightly the u_k's in order to get a smooth function and end the proof of the theorem.

Proof of the Theorem 3.1 : We deal with the operator

$$L = \frac{\partial}{\partial t} + i \sum_{j=1}^{n-1} b_j(x,t) \frac{\partial}{\partial x_j} + c$$

near the origin, with $\sum_{j=1}^{n-1} |b_j(0,0)| \neq 0$.

Step 1 : Existence of ϕ and ξ .

Let δ be a positive number (which stands for b_k), $\theta \in \mathbb{N}$, $\theta > 1$ and $\lambda = \delta^{-\theta}$. Since we are looking for functions u_k near b_k we set $t - \delta = \frac{s}{\lambda}$. The operator L becomes

$$\tilde{L} = \lambda \frac{\partial}{\partial s} + i \sum_{j=1}^{n-1} b_j(x, \delta + \frac{s}{\lambda}) \frac{\partial}{\partial x_j} + c(x, \delta + \frac{s}{\lambda}).$$

Let us define

$$(3.1) \qquad u(x,s,\delta) = e^{i\tau\xi(x,\delta)} e^{\nu\phi(x,s,\delta)} e^{-\gamma(x,\delta)} w(x,s,\delta)$$

where ξ, ϕ and γ are real functions and τ, ν positive parameters. Consider

$$I = e^{-i\tau\xi} e^{-\nu\phi} e^{\gamma} \tilde{L}(e^{i\tau\xi} e^{\nu\phi} e^{-\gamma} w).$$

It is easy to see that

$$(3.2)\quad I = \{\lambda\nu \frac{\partial\phi}{\partial s} - \tau \sum_{j=1}^{n-1} b_j(x,\delta + \frac{s}{\lambda}) \frac{\partial\xi}{\partial x_j}\} w + \lambda \frac{\partial w}{\partial s} +$$

$$+ i\nu(Q\phi)w - i(Q\gamma)w + iQw$$

where

$$Q = \sum_{j=1}^{n-1} b_j(x,\delta + \frac{s}{\lambda}) \frac{\partial}{\partial x_j}.$$

We choose first

$$(3.3)\qquad \tau = \lambda\nu \quad .$$

The first term of the right hand side of (3.2) is equal to

$$I_1 = \tau\{\frac{\partial\phi}{\partial s} - \sum_{j=1}^{n-1} b_j(x,\delta + \frac{s}{\lambda}) \frac{\partial\xi}{\partial x_j}\} w.$$

Lemma 3.2 : There exists a neighborhood ω of $x = 0$, positive constants s_o, δ_o, M and real functions $\xi(x,\delta)$ in $C^\infty(\omega \times]0,\delta_o[)$, $\phi(x,s,\delta)$ in $C^\infty(\omega \times]-s_o,s_o[\times [0,\delta_o[)$ such that

$$(3.4)\qquad \frac{\partial\phi}{\partial s} - \sum_{j=1}^{n-1} b_j(x,\delta + \frac{s}{\lambda}) \frac{\partial\xi}{\partial x_j} = 0$$

$$(3.5)\qquad |\xi(x,\delta)| \leqslant C\ \delta^{-M}\ ,\ (x,\delta) \in \omega \times]0,\delta_o[$$

$$(3.6)\qquad \phi(x,s,\delta) = s - \beta(x,s,\delta).s^2 \text{ with } \beta(0,0,0) > 0.$$

Proof : Let B^∞ be the set of $f = f(x,s,\delta)$ which are C^∞ with respect to (x,s) and satisfy $|D_x^\alpha D_s^\beta f| \leqslant C_{\alpha\beta}$ for all $\alpha \in \mathbb{N}^{n-1}$, $\beta \in \mathbb{N}$, when $\delta \to 0$.

We first show that there exist real functions $B(x,\delta)$, $\xi(x,\delta)$ such that $B(0,0) > 0$ and

$$(3.7)\quad \sum_{j=1}^{n-1} b_j(x,\delta) \frac{\partial\xi}{\partial x_j} = 1 \quad , \quad \sum_{j=1}^{n-1} b_j'(x,\delta) \frac{\partial\xi}{\partial x_j} = \lambda B(x,\delta)$$

where $b'_j = \frac{\partial b_j}{\partial t}$.

Setting $\xi(x,\delta) = \lambda\tilde{\xi}(x,\delta)$, with $\lambda=\delta^{-\theta}$, we write the first equation of (3.7)

$$\sum_{j=1}^{n-1} b_j(x,\delta) \frac{\partial\tilde{\xi}}{\partial x_j} = \delta^{\theta} .$$

Assuming $b_{j_o}(0,0) \neq 0$ we can solve this equation with

$\tilde{\xi}_{|x_{j_o}=0} = \sum_{\ell\neq j_o} x_\ell \xi^o_\ell$ (where ξ^o is given in theorem 3.1)

and

$$\tilde{\xi} \in C^\infty (\omega \times [0,\delta_o[), \; \mathrm{grad}_x \tilde{\xi}(0) = \xi^o.$$

Let us define B by

$$B(x,\delta) = \sum_{j=1}^{n-1} b'_j(x,\delta) \frac{\partial\tilde{\xi}}{\partial x_j}.$$

It follows that (3.7) is satisfied and we have

$B(0,0) = \sum_{j=1}^{n-1} b'_j(0,0)\xi^o \neq 0$ by hypothesis. Taking ξ^o or $-\xi^o$ we get $B(0,0) > 0$. Now $\xi(x,\delta) = \delta^{-\theta}\tilde{\xi}(x,\delta)$ so (3.5) is satisfied.

Now using (3.7) and a Taylor expansion of b_j, equation (3.4) becomes

$$(3.8) \quad \frac{\partial\phi}{\partial s} - 1 + sB + \frac{s^2}{\lambda^2} \sum_{j=1}^{n-1} \int_0^1 b''_j(x,\delta+u.\tfrac{s}{\lambda})\,du.\lambda \frac{\partial\tilde{\xi}}{\partial x_j} = 0.$$

It is easy to see that the coefficient of s^2 belongs to B^∞. Solving this equation with $\phi_{|s=0}=0$ we get a solution of (3.5) satisfying (3.7).

Step 2 : Fixing the parameters.

Let ε be in $]0,\frac{1}{2}[$, put $\theta = 1+\varepsilon$, $\rho = \frac{1}{2\varepsilon}$ and take $\sigma \in]1,\rho[$.

For k big enough let us take

$$\delta = b_k = k^{-\rho} \ , \ \nu_k = k^{\sigma} \ , \ \lambda_k = \delta^{-\theta} = k^{\rho\theta} \ , \ \tau_k = \lambda_k \cdot \nu_k \ .$$

However in the following, we shall sometimes write, for simplicity, $\lambda, \delta, \theta, \nu$, etc...

Step 3 : The choice of γ.

For $t \in [b_{k+1}, b_{k-1}]$ we define

$$(3.9) \quad G_k(x,t) = \nu_k \ \phi(x, \frac{t-b_k}{b_k^\theta}, b_k) - \nu_{k+1} \ \phi(x, \frac{t-b_{k+1}}{b_{k+1}^\theta}, b_{k+1}).$$

Lemma 3.3 : Let us set $m_k = \frac{1}{3} b_k + \frac{2}{3} b_{k+1}$ and $I_k(x) = G_k(x, m_k)$.

Then for $k \longrightarrow \infty$

$$(3.10) \qquad I_k(x) \sim -\rho \, k^{\sigma - \frac{1}{2}} \ .$$

Proof : By (3.6) we have

$$I_k(x) = \nu_k\{p_k - \beta(x,p_k,b_k)p_k^2\} - \nu_{k+1}\{q_{k+1} - \beta(x, q_{k+1}, b_{k+1})q_{k+1}^2\}$$

where

$$p_k = \frac{-2\ell_k}{3 \ b_k^\theta}, \ q_k = \frac{\ell_k}{3 \ b_k^\theta}, \ \ell_k = b_k - b_{k+1}.$$

It follows that

$$I_k(x) = -\frac{1}{3} \ell_k \{2 \frac{\nu_k}{b_k^\theta} + \frac{\nu_{k+1}}{b_{k+1}^\theta}\} + \{\nu_{k+1} \ \beta(\ldots)q_{k+1}^2 - \nu_k \ \beta(\ldots)p_k^2\} = ① + ②.$$

Now let us remark that if $f(k) \sim k^a$ when $k \longrightarrow \infty$ then

$$f(k+1) - f(k) \sim a \ k^{a-1} \ .$$

It follows that $\ell_k \sim \rho k^{-\rho-1}$ and $① \sim -\rho k^{-\rho-1+\sigma+\rho\theta}$. We have

$$\sigma + \rho\theta - \rho - 1 = \sigma + \rho(\theta-1) - 1 = \sigma + \frac{1}{2\varepsilon}.\varepsilon - 1 = \sigma - \frac{1}{2}.$$

On the other hand, it is easy to see that $|(2)| \leq C\, k^{\sigma+2\rho\theta-2\rho-2} = C\, k^{\sigma-1}$.

Let us define

$$\gamma_k(x) = - \sum_{j=k_o}^{k-1} I_j(x).$$

We get

$$(3.11) \qquad \begin{cases} \gamma_{k+1}(x) - \gamma_k(x) = - I_k(x) \\ \gamma_k(x) \sim \dfrac{\rho}{\sigma + \frac{1}{2}} k^{\sigma + \frac{1}{2}}, \quad k \longrightarrow +\infty . \end{cases}$$

We can also write

$$\gamma_k(x) \sim \frac{\rho}{\sigma + \frac{1}{2}} \delta^{-d} \quad \text{where} \quad \delta = b_k = k^{-\rho} \quad \text{and} \quad d = \frac{\sigma + \frac{1}{2}}{\rho}.$$

Let $\tilde{\gamma}(x,\delta)$ be a function in B^∞ such that, with $\gamma(x,\delta) = \delta^{-d}\tilde{\gamma}(x,\delta)$, we have $\gamma(x,b_k) = \gamma_k(x)$. By (3.11), $\tilde{\gamma}(x,0) = \dfrac{\rho}{\sigma + \frac{1}{2}}$.

Step 4 : The transport equations.

By (3.2) and (3.4) we see that

$$I = \lambda\{\frac{\partial w}{\partial s} + i\delta^{\theta}\nu(Q\phi)w - i\delta^{\theta-d}(Q\tilde{\gamma}).w + i\delta^{\theta}\, Qw\}$$

with

$$Q = \sum_{j=1}^{n-1} b_j(x, \delta + \frac{s}{\lambda}) \frac{\partial}{\partial x_j}.$$

Now $\delta^\theta \nu = \delta^\theta \ \delta^{-\sigma/\rho}$ and $\theta - \frac{\sigma}{\rho} > 0$ since $\theta > 1$ and $\frac{\sigma}{\rho} < 1$. On the other hand, $\theta - d = \theta - \frac{\sigma + 1/2}{\rho} = 1 - \frac{\sigma}{\rho} > 0$.

It follows that the operators

$$\begin{cases} L_o = \frac{\partial}{\partial s} + i\delta^\theta \nu(Q\phi) - i\delta^{\theta - d}(Q\tilde{\gamma}) \\ L_1 = iQ \end{cases}$$

have coefficients in B^∞ and

$$I = \lambda(L_o w + \delta^\theta L_1 w).$$

Let $(w_j)_{j \in \mathbb{N}}$ be the solutions in B^∞ of the Cauchy problems

$$(3.12) \qquad \begin{cases} L_o w_o = 0 \\ w_o|_{s=0} = 1 \end{cases} \quad \text{and} \quad \begin{cases} L_o w_j = -L_1 w_{j-1} & j \geq 1 \\ w_j|_{s=0} = 0 \end{cases} .$$

There exists a function $f(x,s,\delta,\xi)$ in B^∞ with respect to (x,s,δ) near the origin $(\xi \in \mathbb{R})$, such that

(3.13) $\forall \ell \in \mathbb{N}$, $\forall \alpha \in \mathbb{N}^{n-1}$, $\forall N > 0 \ \exists \ C = C_{\ell,\alpha,N} > 0$:

$$|D_s^\ell D_x^\alpha (f - \sum_{j=0}^{N-1} \xi^j w_j)| \leq C\, |\xi|^N \quad .$$

Such a function can be easily constructed taking

$$f(x,s,\delta,\xi) = \sum_{j=0}^{\infty} \phi(\lambda_j \xi) \xi^j \ w_j(x,s,\delta)$$

where $\phi \in \mathcal{D}(]-1,1[)$ and (λ_j) is a sequence of real numbers tending to infinity fast enough. We define

$$w_k(x,s) = f(x,s,b_k,b_k^\theta)$$

and for any (x,t) in $\omega \times [b_{k+1}, b_{k-1}]$ we set

$$(3.14)\quad v_k(x,t) = e^{-\gamma_k(x)} e^{i\tau_k \xi(x,b_k)} e^{\nu_k \phi(x, \frac{t-b_k}{b_k^\theta}, b_k)} w_k(x, \frac{t-b_k}{b_k^\theta}).$$

We then have the following

<u>Proposition 3.4</u> : <u>In</u> $\omega \times [b_{k+1}, b_{k-1}]$ <u>we define</u> $f_k = \frac{Pv_k}{v_k}$.
<u>Then for any</u> $\alpha \in \mathbb{N}^{n-1}$, $\ell \in \mathbb{N}$ <u>there exists</u> $\mathcal{H} > 0$ <u>such that for every</u> $N \in \mathbb{N}$ <u>there exists</u> $C = C_{\alpha,\ell,N}$ <u>satisfying</u>

$$(3.15)\qquad |D_x^\alpha D_t^\ell f_k| \leq C\, k^{-\mathcal{H}N}$$

<u>for</u> $(x,t) \in \omega \times (b_{k+1}, b_{k-1})$ <u>and</u> $k \geq k_o$.

<u>Proof</u> : It is easy to see that

$$f_k = b_k^{-\theta} \frac{1}{w_k} \{L_o\, w_k + b_k^\theta\, L_1\, w_k\}.$$

On the other hand, by (3.13),

$$w_k(x,s) = \sum_{j=0}^{N-1} b_k^{j\theta}\, w_j(x,s,b_k) + b_k^{N\theta}\, R_N(x,s,b_k)$$

whith $R_N \in B^\infty$.

Since the (w_j) satisfy (3.12) we have

$$f_k = b_k^{-\theta} \frac{1}{w_k} \{ b_k^{N\theta}\, L_o\, R_N + b_k^{(N+1)\theta} L_1\, R_N + b_k^{N\theta}\, L_1\, w_{N-1}\}$$

from which we deduce

$$|f_k| \leq C_N\, k^{-\frac{1}{2}\rho\,\theta . N}$$

The same argument applies to the derivatives of f_k.

<u>Step 5</u> :

For (x,t) in $\omega \times [b_{k+1}, b_{k-1}]$ let us define

(3.16) $$F_k(x,t) = \mathrm{Log}\left|\frac{v_k(x,t)}{v_{k+1}(x,t)}\right| .$$

The main point in this section is the following

Proposition 3.5 : There exist positive constants C and η with $\eta > \rho+1$ such that

(3.17) $$\frac{\partial F_k}{\partial t}(x,t) \geq C\, k^{\eta}$$

for all (x,t) in $\omega \times [b_{k+1}, b_{k-1}]$ and k big enough.

Proof : We know by (3.6) that $\frac{\partial\phi}{\partial s} = 1 - s\,\beta_1(x,s,\delta)$ with $\beta_1(0,0,0) > 0$. Now we have

$$\frac{\partial F_k}{\partial t} = \left\{\frac{\nu_k}{b_k^{\theta}} - \frac{\nu_{k+1}}{b_{k+1}^{\theta}}\right\} +$$

$$+ \left\{\beta_1\left(x, \frac{t-b_{k+1}}{b_{k+1}^{\theta}}, b_{k+1}\right) \cdot \frac{t-b_{k+1}}{b_{k+1}^{2\theta}} - \beta_1\left(x, \frac{t-b_k}{b_k^{\theta}}, b_k\right) \cdot \frac{t-b_k}{b_k^{2\theta}}\right\} +$$

$$+ \left\{\frac{1}{b_k^{\theta}}\,\frac{w'_k}{w_k} - \frac{1}{b_{k+1}^{\theta}} \cdot \frac{w'_{k+1}}{w_k}\right\} = \textcircled{1} + \textcircled{2} + \textcircled{3} ,$$

we have

$$\textcircled{1} \sim k^{\sigma+\rho\theta-1} \quad \text{when} \quad k \to +\infty .$$

On the other hand, with $t_k = \frac{t-b_k}{b_k^{\theta}}$, we have

$$\textcircled{2} = \beta_1(x, t_{k+1}, b_{k+1}) \left[\left(\frac{t-b_{k+1}}{b_{k+1}^{2\theta}} - \frac{t-b_k}{b_{k+1}^{2\theta}}\right)\nu_{k+1} + (t-b_k)\left(\frac{\nu_{k+1}}{b_{k+1}^{2\theta}} - \frac{\nu_k}{b_k^{2\theta}}\right)\right]$$

$$+ \frac{\nu_k}{b_k^{2\theta}}(t-b_k)\left[\beta_1(x,t_{k+1},b_{k+1}) - \beta_1(x,t_k,b_k)\right] .$$

Since $\beta_1(0,0,0) > 0$ we have, with k big enough,

$$\beta_1(x,t_{k+1},b_{k+1})\,\frac{b_k-b_{k+1}}{b_{k+1}^{2\theta}}\cdot\nu_{k+1}\geqslant C_o\;k^{-\rho-1+2\rho\theta+\sigma}.$$

Now
$$\left|\beta_1(\ldots)(t-b_k)\left(\frac{\nu_{k+1}}{b_{k+1}^{2\theta}}-\frac{\nu_k}{b_k^{2\theta}}\right)\right|\leqslant C_1\,k^{-\rho-1+\sigma+2\rho\theta-1}$$

and

$$\frac{\nu_k}{b_k^{2\theta}}\,(t-b_k)\,[\beta_1(\ldots)-\beta_1(\ldots)]=o(k^{\sigma+2\rho\theta-\rho-1}).$$

It follows that

$$②\geqslant C\;k^{\sigma+2\rho\theta-\rho-1}.$$

It is easy to see that

$$|③|\leqslant C'k^{\theta\rho}.$$

This proves (3.13) with $\eta=\sigma+2\rho\theta-\rho-1$. Now we have $\eta>\rho+1$, since it is equivalent to $\sigma-1>0$.
By (3.9), (3.14) and (3.16) we can write

$$F_k(x,t)=\gamma_{k+1}(x)-\gamma_k(x)+G_k(x,t)+\mathrm{Log}\left|\frac{w_k}{w_{k+1}}\right|.$$

Using (3.11) we get with, $m_k=\frac{1}{3}\,b_k+\frac{2}{3}\,b_{k+1}$,

$$F_K(x,m_k)=\gamma_{k+1}(x)-\gamma_k(x)+I_k(x)+\mathrm{Log}\left|\frac{w_k}{w_{k+1}}\right|=\mathrm{Log}\left|\frac{w_k}{w_{k+1}}\right|=0(1).$$

By (3.17) it is easy to see that in an interval, containing m_k, of length $\frac{c}{k^\eta}$, $F_k(x,t)$ must vanish. It follows from the implicit function theorem that there exists a smooth function $m_k(x)$ such that $F_k(x,m_k(x))=0$. Moreover

$$(3.18)\qquad m_k(x)-m_k=\frac{|F_k(x,m_k(x))-F_k(x,m_k)|}{F_k'(x,\xi_k(x))}\leqslant\frac{C}{k^\eta},$$

and since $\eta > \rho + 1$, the function $m_k(x)$ stays inside $[b_{k+1}, b_k]$.

Step 6 : Modification of v_k.

We are now going to modify slightly the v_k's in order to get a well defined perturbation a. Let us define

$$u_k(x,t) = e^{-\gamma_k(x)} e^{i\tau_k \xi(x,b_k)} e^{\nu_k \phi(x,t_k,b_k)} \{w_k(x,t_k) + z_k(x,t_k)\}$$

where $t_k = \dfrac{t-b_k}{b_k^\theta}$.

We now show that there exists z_k such that :

(i) $F_k = \mathrm{Log}\left|\dfrac{u_k}{u_{k+1}}\right|$ satisfies (3.17) and vanishes for $t = m_k(x)$.

(ii) $g_k = \dfrac{Lu_k}{u_k}$ satisfies (3.15).

(iii) g_k is flat on $\{(x,t) : t = m_k(x)\}$ and $\{(x,t) : t = m_{k-1}(x)\}$.

With this choice, roughly speaking, in $\omega \times (b_{k+1}, b_k)$

$$a = \frac{L(u_k+u_{k+1})}{u_k+u_{k+1}} = \frac{g_k\ u_k + g_{k+1} \cdot u_{k+1}}{u_k + u_{k+1}}$$

and the numerator is flat on the set where the denominator vanishes.

Now for $t = m_k(x)$ we have, using (3.18),

$$\frac{t-b_k}{b_k^\theta} = \frac{m_k(x)-m_k+m_k-b_k}{b_k^\theta} = (-\frac{2}{3}\ell_k + 0(k^{-\eta}))b_k^{-\theta} \ ;$$

so

$$(3.19)\quad \begin{cases} \text{For } t=m_k(x), \dfrac{t-b_k}{b_k^\theta} = (-\frac{2}{3}\rho + e_k(x))k^{\rho\theta-\rho-1}, & e_k(x)=o(1) \\ \text{For } t=m_{k-1}(x), \dfrac{t-b_k}{b_k^\theta} = (\frac{1}{3}\rho + e_k'(x))k^{\rho\theta-\rho-1}, & e_k'(x)=o(1). \end{cases}$$

Let us consider the surfaces

$$S_k = \{(x,\tau) \in \omega \times \mathbb{R} : \tau = -\frac{2}{3}\rho + e_k(x)\}$$

$$\Sigma_k = \{(x,\tau) \in \omega \times \mathbb{R} : \tau = -\frac{1}{3}\rho + e_k'(x)\}.$$

Let us take a function $\tilde{z}_k(x,\tau)$ defined on $\omega \times \mathbb{R}$ and consider

$$z_k(x;s) = \tilde{z}_k(x, k^{-\rho\theta+\rho+1} s).$$

We see easily from (3.19) that

$$(3.20)\quad (\frac{\partial}{\partial t})^\ell z_k(x, \frac{t-b_k}{b_k^\theta})\Big|_{t=m_k(x)} = k^{\alpha(\ell)} (\frac{\partial}{\partial \tau})^\ell \tilde{z}_k\Big|_{S_k},$$

$\alpha(\ell) \in \mathbb{R}$,

and the same for $t = m_{k-1}$ and Σ_k.

Now we have

$$(3.21)\quad \frac{Lu_k}{u_k} = g_k = \frac{w_k}{w_k+z_k} \cdot f_k + \frac{Lz_k}{w_k+z_k}.$$

where

$$L = \frac{\partial}{\partial t} + i\nu(Q\phi) - i\delta^{-d}(Q\tilde{\gamma}) + iQ, \quad Q = \sum_{j=1}^{n-1} b_j(x,\delta + \frac{s}{\lambda}) \frac{\partial}{\partial x_j}.$$

Choose $\tilde{z}_{k|S_k} = 0$. Then by (3.20) and (3.21)

$$g_k|_{t=m_k(x)} = 0 \Longleftrightarrow \frac{\partial}{\partial\tau} \tilde{z}_k|_{S_k} = -k^{\beta(\ell)}.w_k\, f_k|_{t=m_k(x)} = \phi_{1k}(x),$$

and by (3.15) $\phi_{1k}(x)$ is rapidly decreasing with respect to k. Differentiating (3.21) with respect to t we see easily that

$$(3.22)\quad (\frac{\partial}{\partial t})^j g_k|_{t=m_k(x)} = 0 \quad 1 \leqslant j \leqslant \ell \Longleftrightarrow (\frac{\partial}{\partial\tau})^j \tilde{z}_k|_{S_k} = \phi_{jk} \quad 1 \leqslant j \leqslant \ell+1,$$

where ϕ_{jk} are given and rapidly decreasing with respect to k.
The same thing is true for $t = m_{k-1}(x)$ and Σ_k.
By the Whitney theorem [60], given $(\phi_{jk})_{j \in \mathbb{N}^*}$ and $(\psi_{jk})_{j\in\mathbb{N}^*}$, there exists a smooth function $\tilde{z}_k$ such that

$$\begin{cases} (\frac{\partial}{\partial\tau})^j \tilde{z}_k|_{S_k} = \phi_{jk}, & j \in \mathbb{N}^* \\ (\frac{\partial}{\partial\tau})^j \tilde{z}_k|_{\Sigma_k} = \psi_{jk}, & j \in \mathbb{N}^* \end{cases}$$

and $\tilde{z}_k(x,\tau)$ satisfies, for all N,

$$|D_x^\alpha D_\tau^\ell \tilde{z}_k| \leqslant C_{\alpha\ell N}\, k^{-N}$$

for $(x,\tau) \in \omega \times [-2\rho, 2\rho]$.

Step 7 : The end of the proof.

Let $\chi \in C_o^\infty(\mathbb{R})$ be such that

$$\chi(s) = 1 \quad \text{for} \quad |s| \leqslant \frac{3}{4}, \ \operatorname{supp}\chi \subset [-1,+1], \ 0 \leqslant \chi \leqslant 1.$$

We define

$$u(x,t) = \sum_{k \geqslant k_o} \chi_k(t)\, u_k(x,t)$$

with $\chi_k(t) = \chi(\frac{t-b_k}{\ell_k})$ and

$$a = \frac{Pu}{u}$$

(i) For $t \in [b_k - \frac{3}{4}\ell_k, b_{k+1} + \frac{3}{4}\ell_{k+1}]$:

We have

$$a = \frac{L(u_k+u_{k+1})}{u_k+u_{k+1}} = \frac{g_k u_k + g_{k+1} u_{k+1}}{u_k+u_{k+1}} .$$

For $t \geqslant m_k(x)$ we have $|u_k| \geqslant |u_{k+1}|$ since $\tilde{F}_k$ is increasing.

Now

$$|u_k+u_{k+1}| = |u_k| \, |1 + \frac{u_{k+1}}{u_k}| \geqslant |u_k|(1-|\frac{u_{k+1}}{u_k}|) = |u_k| \, (1-e^{-\tilde{F}_k}).$$

Using the inequalities $1 - e^{-x} \geqslant \frac{x}{4}$ if $x \in [0,2]$, $1 - e^{-x} \geqslant \frac{1}{2}$ if $x \geqslant 1$ and (3.17) we get

$$|u_k+u_{k+1}| \geqslant \begin{cases} C\,k^{\eta}\,(t-m_k(x))\,.\,|u_k| & \text{if } \tilde{F}_k \in [0,2] \\ \frac{1}{2}\,|u_k| & \text{if } \tilde{F}_k \geqslant 1 . \end{cases}$$

Since g_k and g_{k+1} are flat on $t = m_k(x)$, this implies that $a \in C^{\infty}$ and

$$|a| \leqslant \sup \left(\frac{|g_k| + |g_{k+1}|}{Ck^{\eta}|t-m_k(x)|}, 2(|g_k| + |g_{k+1}|)\right) .$$

We use the same argument for $t \leqslant m_k(x)$ and for the derivatives of a .

(ii) For $t \in [b_{k+1}, b_k - \frac{3}{4}\ell_k]$:

Here

$$a = \frac{L(u_{k+1} + \chi_k u_k)}{u_{k+1} + \chi_k u_k} .$$

We are going to prove that

$$|\frac{u_k}{u_{k+1}}| \leqslant e^{-ck^{\eta-\rho-1}} , \quad c>0 . \tag{3.23}$$

Indeed, $\tilde{F}_k(x,t) = (t-m_k(x))\frac{\partial \tilde{F}_k}{\partial t}(x,\xi^*)$, $t \leqslant b_k - \frac{3}{4}\ \ell_k$

and

$$m_k(x) = \frac{1}{3}\ b_k + \frac{2}{3}\ b_{k+1} + O(k^{-\eta}).$$

It follows that $t - m_k(x) \leqslant -\frac{1}{12}\ \ell_k + O(k^{-\eta})$. Using (3.17) we get (3.23). Now we can write, if k is big enough,

$$|u_{k+1} + \chi_k u_k| = |u_{k+1}| \left| 1 + \chi_k \frac{u_k}{u_{k+1}} \right| \geqslant \frac{1}{2}\ |u_{k+1}|.$$

This proves that a is well-defined and

$$|a| \leqslant \frac{2(|g_{k+1}||u_{k+1}| + |[L,\chi_k]|.\ |u_k| + \chi_k\ |g_k||u_k|)}{|u_{k+1}|},$$

so

$$|a| \leqslant 2(|g_k| + |g_{k+1}| + e^{-ck^{\eta-\rho-1}}.\ O(k^M)).$$

which proves that a is rapidly decreasing. The same is true for the derivatives of a.

(iii) For $t \in [b_{k+1} + \frac{3}{4}\ \ell_{k+1},\ b_k]$:

We use exactly the same argument as in (ii).
Now the function u is C^∞ and flat on $t=0$, since by (3.11),

$$\gamma_k(x) \geqslant c\ k^{\sigma+\frac{1}{2}} \quad \text{and} \quad \nu_k\phi(\ldots) = O(k^\sigma).$$

Moreover u vanishes, for $t>0$, at most on $\{t = m_k(x)\}$, which proves that the origin is in the support of u. This ends the proof of theorem 3.1. ∎
Let us remark that with theorems 2.1 and 3.1 we have a quite complete answer to the problem of uniqueness for first order differential operators, at least when the dimension n is greater or equal to 3 (indeed Theorem 3.1 is empty when $n=2$). We are now going to look at the operators in $\mathbb{R}^2$.

§ 4. THE TWO DIMENSIONAL CASE

We begin by stating the following

Theorem 4.1 : *Let us consider, in a neighborhood* V *of the origin in* $\mathbb{R}^2$, *the operator*

$$L = \frac{\partial}{\partial t} + ib(x,t)\frac{\partial}{\partial x} + c$$

where b *and* c *are* C^∞ *in* V *and* b *is real. Let us suppose that* $t \mapsto b(0,t)$ *has at* $t=0$ *a zero of finite order. Then there exists a neighborhood* W *of the origin in which every* $u \in C^1(V)$, *satisfying* $Lu = 0$ *and* $u|_{t<0} = 0$, *vanishes.*

This result proves that we can get uniqueness even without condition (P) .

Proof :

Step 1 :

It is sufficient to prove the above result for the operator

$$L = \frac{\partial}{\partial t} + (\alpha+i\beta)t^\ell \frac{\partial}{\partial x} + \gamma, \ \beta(0,0) \neq 0, \ \ell \in \mathbb{N}. \tag{4.1}$$

Indeed, by the Malgrange-Weierstrass Theorem, we can write for $|x| < r$ and $0 < t < T$,

$$b(x,t) = a(x,t)\ (t^k + a_1(x)t^{k-1} + \ldots + a_k(x))$$

where a and a_j are smooth, $a(0,0) \neq 0$ and $a_j(0) = 0$. Let us set $f = \frac{b}{a}$. We have

Lemma 4.2 :

a) *There exists* k *open sets* $\mathcal{O}_1, \ldots \mathcal{O}_k$ *contained in* $|x| < r$ *whose union is dense, such that for* $j=1,\ldots,k$ *and* $x \in \mathcal{O}_j$, $t \to f(x,t)$ *has exactly* j *distinct roots.*

b) *In each* $\mathcal{O}_j$ *the distinct roots of* f *can be represented by* j C^∞ *functions* $\rho_i(x)$.

c) <u>For</u> $j=1,\dots,k$, <u>and</u> $i=1,\dots,j$ <u>if</u> $\rho_i(x)$ <u>is real at</u> $x_o \in \mathcal{O}_j$, <u>it is real in the connected component of</u> x_o <u>in</u> $\mathcal{O}_j$.

<u>Proof</u> : For $x \in]-r, r[$ let $d(x)$ be the number of distinct roots of $t \to f(x,t)$; then $d(x)$ is upper semi-continuous with values in $\{1,2,\dots,k\}$. The set where $d(x) = k$ is then open, let us call it $\mathcal{O}_k$. Define $\mathcal{O}'_k = \overbrace{]-r,r[\setminus \mathcal{O}_k}^{o}$ and put

$$\mathcal{O}_{k-1} = \{x \in \mathcal{O}'_k : d(x) = k-1\}$$

and so on. We get, by induction, open sets $\mathcal{O}_1,\dots,\mathcal{O}_k$ which satisfy a).
The part b) follows from the implicit function theorem. To get c), let $\rho(x)$ be, in $\mathcal{O}_j$, a root of f ; suppose $\rho(x_o)$ real and $\rho(x)$ non-real in a neighborhood of x_o. Then $\bar{\rho}(x)$ is a root of f, different from ρ in $\mathcal{O}_j$ such that $\rho(x_o) = \bar{\rho}(x_o)$, which is impossible.■

Let $u \in C^1(V)$ be the solution defined in theorem 4.1. Suppose $F = \operatorname{supp} u \neq \phi$. Then there exists j and a connected component U of $\mathcal{O}_j$ such that $F \cap U \times]0,T[\neq \phi$. Let $\rho_1(x) < \dots < \rho_\ell(x)$, $\ell \leq j$, be the real roots of f in U. Define $\tilde{\rho}_o(x) = 0$, $\tilde{\rho}_i(x) = \sup(0, \inf(\rho_i(x),T))$ for $1 \leq i \leq \ell$ and $\tilde{\rho}_{\ell+1}(x) = T$.
Let us define for $0 \leq i \leq \ell$,

$$A_i = \{(x,t) \in U \times]0,T[: \tilde{\rho}_i(x) \leq t \leq \tilde{\rho}_{i+1}(x)\}$$

and $i_o = \operatorname{Inf}\{i : F \cap A_i \neq \phi\}$.

a) Suppose there exists $(x_o,t_o) \in F \cap A_{i_o}$ with $t_o = \rho_{i_o}(x_o)$. We perform the following change of coordinates: $x' = x - x_o$, $t' = t - \rho_{i_o}(x)$. In these coordinates near (x_o,t_o) the operator L becomes

$$\tilde{L} = \frac{\partial}{\partial t'} + it'^{\ell}(\alpha+i\beta)\frac{\partial}{\partial x'} + \gamma$$

and the new solution $\tilde{u}$ vanishes for $t'<0$. It follows from (4.1) that $\tilde{u}$ vanishes near the origin, which means that u vanishes near (x_o,t_o), in contradiction with a).

b) Suppose $F \cap A_{i_o} \cap \{t = \rho_{i_o}(x)\} = \phi$.

Let x_o be the mid-point of U and perform the same change of coordinates as before. The set $\{(x,t) \in U \times]0,T[: t = \rho_{i_o}(x)\}$ is transformed into $\{(x',t') : t' = 0, |x'| < a\}$. Let

$$\varepsilon_o = \mathrm{Min}\{\varepsilon > 0 : \{\text{Graph of } t' - \varepsilon = \frac{-\varepsilon x'^2}{a^2}\} \cap \tilde{A}_{i_o} \cap \mathrm{supp}\,\tilde{u} \neq \phi\}.$$

By hypothesis $\varepsilon_o > 0$. Take $(x'_o,t'_o) \in \tilde{A}_{i_o} \cap \mathrm{supp}\,\tilde{u}$ such that

$$t'_o = -\frac{\varepsilon_o}{a^2}x'^2_o + \varepsilon_o .$$

Near this point the operator L is elliptic, since b does not vanish there and the support of the solution $\tilde{u}$ is from one side of the parabola $t' = -\frac{\varepsilon_o}{a^2}x'^2 + \varepsilon_o$.

We conclude that $\tilde{u}$ must vanish near this point, which is a contradiction.

<u>Step 2</u> : <u>The uniqueness for the operator (4.1)</u>

We prove a slightly stronger version of the uniqueness result that we need.

<u>Proposition 4.2</u> : <u>Let us consider, in a neighborhood</u> V <u>of the origin in</u> $\mathbb{R}^2$, <u>the operator</u>

(4.2) $$L = \frac{\partial}{\partial t} + (f+ig)t^{\ell}\frac{\partial}{\partial x} + \frac{h}{t}$$

<u>where</u> $f,g \in C^{\infty}(V)$, $g(0,0) \neq 0$, $h \in L^{\infty}(V)$.

<u>There exists a neighborhood of the origin in which every</u> $u \in C^\infty(V)$ <u>satisfying</u> $Lu = 0$ <u>in</u> V <u>and</u> $u_{|t<0} = 0$ <u>vanishes.</u>

<u>Proof</u> : Let us perform near the origin the following singular change of coordinates

$$X = x,\ t = (\delta - X^2)T,\ \delta > 0 \text{ small}.$$

It is easy to see that operator (4.2) transforms to $\tilde{L}$ such that

$$(\delta - X^2)\tilde{L} = A\frac{\partial}{\partial T} + T^\ell(\delta - X^2)^{\ell+1}(B+iC)\frac{\partial}{\partial X} + \frac{D}{T}$$

where

$$\begin{cases} A,B,C \in C^\infty\ ,\ D \in L^\infty \\ A(0,0) = 1,\ C(0,0) \neq 0. \end{cases}$$

Dividing by A and writing for simplicity (x,t) instead of (X,T), we see that it is sufficient to deal with the operator

$$P = \frac{\partial}{\partial t} + t^\ell(\delta - x^2)^{\ell+1}(a+ib)\frac{\partial}{\partial x} + \frac{\alpha}{t} \tag{4.3}$$

with $b(0,0) \neq 0$ and the support of the solution u contained in the set $\{(x,t) : t \geq 0,\ |x|^2 \leq \delta\}$. We state now the

<u>Proposition 4.3</u> : <u>There exist positive constants</u> C, γ_0, T_0, r <u>such that for</u> $\gamma \geq \gamma_0$ <u>and every</u> $v \in C^\infty$ <u>with</u>

$$\operatorname{supp} v \subset \{0 \leq t \leq T_0,\ |x| \leq r\},$$

<u>we have</u>

$$\gamma\, \|t^{-\gamma-1} v\|^2 \leq C\, \|t^{-\gamma} Pv\|^2, \tag{4.4}$$

<u>where</u> $\|\ \|$ <u>stands for the</u> L^2 <u>norm.</u>

The uniqueness for P follows from Proposition 4.3 by a standard argument (see the proof of Theorem 1.1).

Proof : Let us set $v = t^{\gamma}w$ and $P_o = \frac{\partial}{\partial t} + (a+ib)t^{\ell}(\delta-x^2)^{\ell+1}\frac{\partial}{\partial x}$.
It follows that $t^{-\gamma}P_o v = Xw + Yw$ with

$$(4.5)\qquad \begin{cases} X = \frac{\partial}{\partial t} + (\delta-x^2)^{\ell+1}\, t^{\ell}\, a\, \frac{\partial}{\partial x} \\ Y = \gamma t^{-1} + i(\delta-x^2)^{\ell+1}\, t^{\ell}\, b\, \frac{\partial}{\partial x}. \end{cases}$$

Now

$$(4.6)\qquad \| t^{-\gamma}P_o v\|^2 = \|Xw\|^2 + \|Yw\|^2 + 2\mathrm{Re}(Xw, Yw).$$

It is easy to see that

$$(4.7)\qquad \begin{cases} 2\mathrm{Re}(\partial_t w, \gamma t^{-1}w) = \gamma\, \|t^{-1}w\|^2 \\ |2\mathrm{Re}((\delta-x^2)^{\ell+1}t^{\ell}\, a\,\partial_x w, \gamma t^{-1}w)| \leqslant C\|t^{\frac{\ell-1}{2}}w\|^2 \\ 2\mathrm{Re}((\delta-x^2)^{\ell+1}\, t^{\ell}\, a\, \partial_x w,\ i(\delta-x^2)^{\ell+1}\, t^{\ell}\, b\, \partial_x w) = 0. \end{cases}$$

On the other hand,

$$2\mathrm{Re}(\partial_t w,\ i(\delta-x^2)^{\ell+1}\, t^{\ell}\, b\, \partial_x w) = \underset{1/}{-(w,\ i(\delta-x^2)^{\ell+1}\ell t^{\ell-1} b\partial_x w)} -$$

$$\underset{2/}{-(w,\ i(\delta-x^2)^{\ell+1}\, t^{\ell}\, b'_t\, \partial_x w)} + \underset{3/}{(w,\ i(\delta-x^2)^{\ell+1}\, t^{\ell}\, b'_x\, \partial_t w)} -$$

$$\underset{4/}{-(w, i(\ell+1)2x\ (\delta-x^2)^{\ell}\ t^{\ell}\ b\ \partial_t w)}.$$

We have using (4.5),

$$1/ = -\ell(t^{-1}w,\ i(\delta-x^2)^{\ell+1}\, t^{\ell}\, b\, \partial_x w) = -\ell(t^{-1}w, Yw) + \ell(t^{-1}w, \gamma t^{-1}w).$$

It follows that for every $\varepsilon > 0$ we can find $c_{\varepsilon} > 0$ such that

$$(4.8)\qquad \begin{cases} 1/ = \ell\,\gamma\, \|t^{-1}w\|^2 + R \\ |R| \leqslant \varepsilon\, \|Yw\|^2 + c_{\varepsilon}\, \|t^{-1}w\|^2. \end{cases}$$

Now, since $b(0,0) \neq 0$, we can write

$$2/ = -(\frac{b_t'}{b} w, \; i(\delta - x^2)^{\ell+1} t^{\ell} b \partial_x w) = -(\frac{b_t'}{b} w, Yw) + (\frac{b_t'}{b} w, t^{-1} w)$$

from which we deduce

$$(4.9) \qquad |2/| \leqslant \varepsilon \|Yw\|^2 + c_\varepsilon \|w\|^2 + c\gamma \|t^{-1/2} w\|^2 .$$

Using (4.5) we get, with $f = (\delta - x^2)^{\ell+1} t^{\ell} b_x'$,

$$3/ = (fw, \; iXw) - (fw, \; i(\delta - x^2)^{\ell+1} t^{\ell} a \, \partial_x w) =$$

$$= (fw, \; iXw) - (\frac{f}{b} w, \; i(\delta - x^2)^{\ell+1} t^{\ell} b \, \partial_x w).$$

By the same argument as before we deduce

$$(4.10) \quad |3/| \leqslant \varepsilon(\|Xw\|^2 + \|Yw\|^2) + c_\varepsilon \|w\|^2 + c\gamma \|t^{-1/2} w\|^2 .$$

Finally

$$4/ = -(w, i(\ell+1) 2x \; (\delta - x^2)^{\ell} \; t^{\ell} \; bXw) +$$

$$+ (w, i(\ell+1) 2x \; (\delta - x^2)^{2\ell+1} \; t^{2\ell} \; a \, \partial_x w).$$

So we can use the same argument, and get for 4/ the estimate (4.10). By (4.6)...(4.10) we get, with another ε,

$$\|Xw\|^2 + \|Yw\|^2 + (\ell+1)\gamma \|t^{-1} w\|^2 \leqslant \|t^{-\gamma} P_o v\|^2$$

$$+ \varepsilon(\|Xw\|^2 + \|Yw\|^2) + c_\varepsilon \|t^{-1} w\|^2 + c\gamma \|t^{-\frac{1}{2}} w\|^2 .$$

Taking $\varepsilon = \frac{1}{2}$ and T_o^{-1}, γ_o big enough, we deduce that

$$\gamma \|t^{-\gamma-1} v\|^2 \leqslant c_o \|t^{-\gamma} P_o v\|^2 .$$

Since $P = P_o + \alpha t^{-1}$ with $\alpha \in L^\infty$, it follows, increasing γ_o if necessary, that (4.4) holds ∎

<u>Remark 4.4</u> : The conclusion of Proposition 4.2 is false for C^1 functions u satisfying $u_{|t=0} = 0$ without any condition on h, as it is shown by the following example :

$$(\frac{\partial}{\partial t} + (f + ig)t^{\ell} \frac{\partial}{\partial x} - \frac{\alpha}{t})t^{\alpha} H(t) \otimes 1_x = 0$$

where $H(t)$ is Heaviside function and $\alpha \geqslant 2$.
However we can show, using $e^{-\gamma t}$ as weight function, that uniqueness holds if $h(0,0) = 0$.■

Now, even in the two dimensional case, we cannot expect to get uniqueness without any hypothesis on the principal part, as shown in the following

Theorem 4.5 : *There exist two functions a and u in $C^{\infty}(\mathbb{R}^2)$, vanishing for $t < 0$, such that*

$$(4.11) \quad \begin{cases} \frac{\partial u}{\partial t} + a \frac{\partial u}{\partial x} = 0 \quad \text{in} \quad \mathbb{R}^2 \\ \operatorname{supp} u = \{(x,t) : t \geqslant 0\}. \end{cases}$$

Proof : Let $\theta_1 \in C^{\infty}(\mathbb{R})$ be an increasing function such that

$\theta_1(t) = -\frac{1}{2}$ if $t < -\frac{1}{4}$, $\theta_1(t) = 0$ if $|t| \leqslant \frac{1}{8}$, $\theta_1(t) = 1$ if $t > \frac{1}{4}$.

Let $\theta_2(t) \in C^{\infty}(\mathbb{R})$ be an increasing function of $|t|$ such that

$$\theta_2(t) = 1 \text{ if } |t| > \frac{1}{4}, \ \theta_2(t) = 0 \text{ if } |t| < \frac{1}{8}.$$

For $k \geqslant 1$ we set $b_k = \sum_{j=k}^{\infty} \frac{1}{j^2}$; then $b_k - b_{k+1} = \frac{1}{k^2}$. Let us consider

$$(4.12) \quad f_k(t) = -k^2 - k^4(t-b_k)\,\theta_1(k^2(t-b_k)).$$

It is a C^{∞} function which is linear for $t - b_k < -\frac{1}{4k^2}$ with slope $\frac{1}{2}k^4$, and for $t-b_k > \frac{1}{4k^2}$ with slope $-k^4$.

We consider

$$(4.13) \qquad u_k(x,t) = e^{i\lambda_k x} e^{f_k(t)}$$

where λ_k are real numbers to be chosen, and we define

$$u(x,t)=\begin{cases} u_{k_o}(x,t) & \text{if } t \geq b_{k_o} \\ u_k \cdot \theta_2(k^2(t-b_{k+1})) + u_{k+1} \cdot \theta_2(k^2(t-b_k)) & \text{if } t \in (b_{k+1},b_k),\ k \geq k_o \\ 0 & \text{if } t < 0 . \end{cases}$$

We are going to show that, with a good choice of (λ_k) and k_o, the function

$$a(x,t)=\begin{cases} -\dfrac{\frac{\partial u}{\partial t}(x,t)}{\frac{\partial u}{\partial x}(x,t)} & \text{for} \quad t > 0 \\ 0 \text{ for} \quad t \leq 0 \end{cases}$$

belongs to $C^\infty(\mathbb{R}^2)$.

Step 1 : $u \in C^\infty(\mathbb{R}^2)$.

Obviously u is C^∞ for $t \geq b_{k_o}$ and $t \in (b_{k+1},b_k)$. Now for t in $\left[b_{k_o} - \frac{1}{8k_o^2}, b_{k_o} + \frac{1}{8k_o^2}\right]$ we have $\theta_2(k_o^2(t-b_{k_o}))$ $= 0$. Moreover,

$$k_o^2|t-b_{k_o+1}| = k_o^2|b_{k_o+1} - b_{k_o} - (t-b_{k_o})| \geq$$

$$\geq k_o^2|b_{k_o} - b_{k_o+1}| - k_o^2|t-b_{k_o}| \geq 1 - \frac{1}{8} > \frac{1}{4} ,$$

so $\theta_2(k_o^2(t-b_{k_o+1})) = 1$. It follows that for

$|t-b_{k_o}| \leq \frac{1}{8k_o^2}$, $u = u_{k_o}$.

Let us prove now that u is flat on $t=0$.
First of all, for $t \in (b_{k+1}, b_k)$, $-k^4(t-b_k)\theta_1\ (k^2(t-b_k)) \leq 0$,
so by (4.12), $f_k(t) \leq -k^2$.
Now it is easy to see that

$$|D_x^\alpha\ D_t^\beta\ u| \leq c_{\alpha\beta}\ \lambda_{k+1}^\alpha\ k^{c.\beta}\ e^{-k^2}.$$

Let us chose (λ_k) such that

$$(4.14) \qquad |\lambda_k| \leq \lambda^k\ ,\ \lambda > 1\ .$$

It follows that

$$|D_x^\alpha\ D_t^\beta\ u| \leq c'_{\alpha\beta}\ \lambda^{k\alpha}\ k^{c\beta}\ e^{-k^2}\ ,$$

which proves our claim.

Step 2 : $a \in C^\infty(\mathbb{R}^2)$.

We divide $[b_{k+1}, b_k]$ in three parts, namely

$$I_k^1 = \{t : b_{k+1} \leq t \leq b_{k+1} + \frac{1}{4k^2}\}$$

$$I_k^2 = \{t : b_{k+1} + \frac{1}{4k^2} \leq t \leq b_k - \frac{1}{4k^2}\}$$

$$I_k^3 = \{t : b_k - \frac{1}{4k^2} \leq t \leq b_k\}\ .$$

(i) For $t \in I_k^2$:
Here $t-b_k \leq -\frac{1}{4k^2}$ so $\theta_1(k^2(t-b_k)) = -\frac{1}{2}$ and
$f_k = \frac{k^4}{2}(t-b_k) - k^2$.

On the other hand ,

$$f_{k+1}(t) = -(k+1)^2 - (k+1)^4(t-b_{k+1})\theta_1((k+1)^2(t-b_{k+1})).$$

Since $t-b_{k+1} \geqslant \frac{1}{4k^2}$, we have $(k+1)^2(t-b_{k+1}) \geqslant \frac{(k+1)^2}{4k^2} > \frac{1}{4}$, which implies $\theta_1((k+1)^2(t-b_{k+1})) = 1$. It follows that

$$f_{k+1}(t) = -(k+1)^2 - (k+1)^4(t-b_{k+1}).$$

Now $k^2(t-b_{k+1}) \geqslant \frac{1}{4}$ and $k^2(t-b_k) \leqslant -\frac{1}{4}$, so $\theta_2(k^2(t-b_{k+1})) = 1$ and $\theta_2(k^2(t-b_k)) = 1$, which implies that

$$u(x,t) = u_k + u_{k+1} = e^{i\lambda_k x + f_k(t)} + e^{i\lambda_{k+1}x + f_{k+1}(t)}.$$

It is then easy to see that

$$a(x,t) = -\frac{\partial_t u}{\partial_x u} = -\frac{\frac{1}{2}k^4 u_k - (k+1)^4 u_{k+1}}{i\lambda_k u_k + i\lambda_{k+1} u_{k+1}}.$$

Let us take $\lambda_{k+1} = c_k \lambda_k$ where c_k is to be chosen. It follows that

$$a(x,t) = -\frac{\frac{1}{2}k^4\left(u_k - 2\frac{(k+1)^4}{k^4} \cdot u_{k+1}\right)}{i\lambda_k(u_k + c_k u_{k+1})}.$$

Let us choose $c_k = -2\frac{(k+1)^4}{k^4}$; then $a(x,t) = -\frac{k^4}{2i\lambda_k}$.

We must prove that this choice of c_k does not contradict (4.14).

Indeed , $\frac{\lambda_{k+1}}{\lambda_k}$ tends to -2 , which implies that

$$c_o\left(\frac{3}{2}\right)^k \leqslant |\lambda_k| \leqslant [\operatorname{Max}(\lambda_o,3)]^k,$$

so $|a| \leqslant ck^4\left(\frac{2}{3}\right)^k \to 0$.

(ii) <u>For</u> $t \in I_k^1$:

Here $\theta_1((k+1)^2\ (t-b_{k+1})) \leqslant 1$ and $t-b_{k+1} \leqslant \frac{1}{4k^2}$. It follows that

$$f_{k+1}(t) \geqslant -(k+1)^2 - \frac{(k+1)^4}{4k^2} = -\frac{5}{4}k^2 + O(k).$$

On the other hand, $t-b_k \leqslant -\frac{1}{4k^2}$ so $\theta_1(k^2(t-b_k)) = -\frac{1}{2}$ and

$$f_k(t) = -k^2 + \frac{1}{2}k^4\,(t-b_k) \leqslant -k^2 - \frac{1}{2}k^4 . \frac{3}{4k^2} = -\frac{11}{8}k^2 .$$

For k big enough ,

(4.15) $$f_{k+1}(t) \geqslant f_k(t) + \frac{k^2}{16} .$$

Now

$$u(x,t) = u_k .\ \theta_2(k^2(t-b_{k+1})) + u_{k+1} ,$$

from which we deduce

$$\frac{\partial u}{\partial x} = i\ \lambda_k\ u_k\ \theta_2 + i\ \lambda_{k+1}\ u_{k+1} =$$

$$= u_{k+1}\ [\,i\ \lambda^k\ e^{i(\lambda_k-\lambda_{k+1})x\ +(f_k-f_{k+1})(t)} + i\ \lambda_{k+1}\ ]$$

$$\frac{\partial u}{\partial x} = i\ \lambda_{k+1}\ u_{k+1}\ [\ 1 + \frac{\lambda_k}{\lambda_{k+1}}\ e^{i(\dots)}\ ] = i\lambda_{k+1}\ u_{k+1}(1+\eta_k).$$

Using (4.14) and (4.15) it is easy to see that

(4.16) $$|D^\alpha\ \eta_k| \leqslant c_\alpha\ \lambda^{k|\alpha|}\ e^{-\frac{k^2}{16}} .$$

Now

$$\frac{\partial u}{\partial t} = \frac{\partial u_k}{\partial t} . \theta_2 + u_k . k^2\ \theta_2' + \frac{\partial u_{k+1}}{\partial t}$$

$$\frac{\partial u}{\partial t} = u_{k+1}\left[f_{k+1}' + g_k(t)e^{i(\lambda_k-\lambda_{k+1})x + f_k(t) - f_{k+1}(t)}\right].$$

Then we can write

$$a(x,t) = -\frac{f'_{k+1} + g_k(t)\, e^{i(\lambda_k - \lambda_{k+1})x + f_k(t) - f_{k+1}(t)}}{i\,\lambda_{k+1}(1 + \eta_k)} .$$

Using (4.15) we deduce easily that

$$|D^\alpha\ a(x,t)| \leqslant c_\alpha\ e^{-\frac{k^2}{16}} \to 0 .$$

(iii) <u>For</u> $t \in I_k^3$:

Here $f_k(t) - f_{k+1}(t) \geqslant c\ k^2$ so we can use the same argument as in (ii). So a is in $C^\infty(\mathbb{R}^2)$ and satisfies (4.11).■

COMMENTS

a) Theorem 2.1 is due to M. Strauss - F. Trêves [51] . The proof given here is simpler than the original one (see Th. 2.2). For an interpretation of the weight function (2.1) used in the Carleman estimate (2.3), we refer to the above paper [51] (see also F. Trêves [53]).

Let us also mention that a different proof of Theorem 2.1, which is not based on Carleman estimates, can be found in M.S Baouendi - F. Trêves [54].

b) The non-uniqueness result (Theorem 3.1) is a particular case of a result of S. Alinhac [1] . The proof given here follows closely that given in S. Alinhac - C. Zuily [5] (see the "Fundamental Lemma").

c) The result of Theorem 4.1 is also due to M. Strauss - F. Trêves [51], with a different proof. The use of the singular change of coordinate in Proposition 4.2 (instead of the convexification) is taken from M.S. Baouendi - E. C. Zachmanoglou [7] .

d) The counterexample given by Theorem 4.1 is due to P. Cohen [12] . The proof given here follows that given by L. Hörmander in his book [19] .

Chapter Two

CALDERON'S THEOREM AND ITS EXTENSIONS

§1. THE CASE OF LOW MULTIPLICITY

In a neighborhood V of a point x_o in $\mathbb{R}^n$ let

$$S = \{x \in V : \phi(x) = \phi(x_o)\}$$

be a C^∞ hypersurface and $P = \sum_{|\alpha| \leq m} a_\alpha(x) D^\alpha$ a differential operator of order m whose principal symbol $p_m(x,\xi)$ has C^∞ coefficients in V.

As usual, the hypersurface S will be assumed to be non characteristic for P at x_o, which means that, with $N_o = d\phi(x_o)$,

(H.1) $$p_m(x_o, N_o) \neq 0 .$$

The main theorem of this section will involve some conditions which we discuss now. The first one is related with the smoothness of the characteristic roots, namely :

(H.2) There exist a conic neighborhood Γ_{N_o} and m functions $\lambda_j(x,\xi,N)$ which are C^∞ in $V \times \mathbb{R}^n \backslash 0 \times \Gamma_{N_o}$ such that for every ξ non-parallel to N,

$$p_m(x,\xi+\tau N) = p_m(x,N) \prod_{j=1}^{m} (\tau - \lambda_j(x,\xi,N))$$

in $V \times \mathbb{R}^n \backslash 0 \times \Gamma_{N_o}$.

On the other hand, we shall need the following :

(H.3) If a root $\lambda_j(x,\xi,N)$ is real (complex) at a point, it remains real (complex) at every point.

The last condition gives a meaning to the title of this section.

(H.4) $\begin{cases} \text{(i) The real roots are simple.} \\ \text{(ii) The multiplicity of the complex roots is at most two.} \end{cases}$

The meaning of (i) in (H.4) is clear. Condition (ii) means the following. Let us denote by $\lambda_{i_1},\ldots,\lambda_{i_p}$ the complex roots; then $\{i_1,\ldots,i_p\} = A\cup B$ with $A\cap B=\phi$ and for every couple (i,j) in A (or in B) we have $\lambda_i(x,\xi,N) \neq \lambda_j(x,\xi,N)$ for every (x,ξ,N) in $V_{x_0} \times \mathbb{R}^n\backslash 0 \times \Gamma_{N_0}$.

We are now able to state the

Theorem 1.1 :

Let P be a differential operator of order m in V, with C^∞ principal part and bounded lower order terms, satisfying conditions (H.1),...,(H.4). Then there exists a neighborhood W of x_0 such that if u is a C^∞ function in V which, with $C = C(u) > 0$, satisfies

$$(1.1)\quad \begin{cases} |Pu(x)| \leq C \displaystyle\sum_{|\alpha|\leq m-1} |D^\alpha u(x)|, \quad x\in V \\ \left(\frac{\partial}{\partial\nu}\right)^j u_{|S} = 0, \quad 0\leq j\leq m-1 \end{cases}$$

then u vanishes in W.

Remark 1.2 :

This theorem contains obviously the uniqueness of the Cauchy problem for P without any conditions on the bounded lower order terms. But it also applies for instance, to non-linear equations. It can be applied to elliptic second order operators with real and C^∞ principal part and also to operators with Δ^2 as principal part.
Let us note that, in Theorem 1.1, the multiplicity of the roots is not assumed to be constant.

The remainder of this section will be devoted to the proof of this result. First, assuming $x_o = 0$, we can, by the classical "Holmgren transformation"

$$x_i \to x_i \quad 1 \leq i \leq n-1, \quad t = \langle x, N_o \rangle + A|x|^2$$

with a suitable constant A, reduce the problem to the following :
In $\mathbb{R}^{n+1} = \{(x,t) : x \in \mathbb{R}^n, t \in \mathbb{R}\}$ we have a differential operator $P(x,t\,;D_x,D_t)$ whose principal symbol can be written as

$$(1.2) \qquad p_m(x,t\,;\xi,\tau) = a(x,t) \prod_{j=1}^{m} (\tau - \lambda_j(x,t\,;\xi)),$$

where $a(0,0) \neq 0$ and $\lambda_j(x,t\,;\xi)$ is a C^∞ function of $(x,t\,;\xi)$ in $V_{(0,0)} \times \mathbb{R}^n\setminus 0$, homogeneous of order 1 with respect to ξ. Moreover $\{\lambda_j\}$ satisfy conditions (H.3) and (H.4). On the other hand, if $\tilde{u}$ denotes the new solution under study, we have

$$\operatorname{supp} \tilde{u} \subset \{(x,t) \in \mathbb{R}^{n+1} : t \geq C|x|^2\}.$$

The result being local near the origin, we can assume that $\lambda_j(x,t\,;\xi)$ vanishes outside the ball $|x|^2 + t^2 \leq k$; indeed, we can write

$$\psi^m(x,t)\, p_m(x,t\,;\xi,\tau) = \psi^m(x,t)\, a(x,t) \prod_{j=1}^{m} (\tau - \phi(x,t)\, \lambda_j(x,t\,;\xi)),$$

where $\phi \in \mathcal{D}(V)$, $\psi \in \mathcal{D}(V_1)$, $\phi = 1$ on V_1, $\psi = 1$ on V_2, $V_2 \subset\subset V_1 \subset\subset V$. The uniqueness result of theorem 1.1 will follow as usual (see the proof of theorem 1.1 Chapter One) from the following:

<u>Theorem 1.3</u> :

<u>There exist positive constants</u> C, T_o, k_o, r <u>such that for</u> $T \leq T_o$, $k \geq k_o$, <u>we have</u>

$$(1.3)\quad \begin{aligned} &\sum_{|\alpha|\leqslant m-1} \int_0^T e^{k(t-T)^2} \|D^\alpha v\|^2\,dt \\ &\leqslant C\left(\frac{1}{k}+T^2\right)\int_0^T e^{k(t-T)^2}\|Pv\|^2\,dt \end{aligned}$$

<u>for every</u> $v\in C^\infty$ <u>with</u> supp $v\subset\{(x,t) : 0\leqslant t\leqslant T,\ |x|\leqslant r\}$. <u>Here</u> $\|\ \|$ <u>stands for the</u> $L^2(\mathbb{R}^n)$ <u>norm</u>.

Before we start the proof of this result, let us introduce some notation .

a) For $\ell\in\mathbb{N}$ let us set

$$|||u|||_\ell^2 = \sum_{|\alpha|+q\leqslant\ell}\int_0^T e^{k(t-T)^2}\|D_t^q D_x^\alpha u\|^2_{L^2(\mathbb{R}^n)}\,dt\ .$$

b) For $r\in\mathbb{N}$ we denote by $S^r(\mathbb{R}^n)$ the set of symbols on $\mathbb{R}^{2n}$ of order r; i.e. the set of C^∞ functions $a(x,\xi)$ on $\mathbb{R}^{2n}$ such that for every α,β in $\mathbb{N}^n$,

$$(1.4)\qquad \sup_{(x,\xi)\in\mathbb{R}^{2n}} |D_x^\alpha D_\xi^\beta a(x,\xi)|(1+|\xi|)^{-r+|\beta|}\leqslant C_{\alpha\beta}\ .$$

Now $T(r)$ will denote the set of functions $a(x,t\,;\xi)$ such that $t\to a(x,t;\xi)$ is a C^∞ function of $t\in[0,T_0]$ with values in $S^r(\mathbb{R}^n)$, with the topology defined by (1.4). OPT(r) will be the set of pseudo-differential operators (p.d.o) with symbols in $T(r)$.

c) Let us recall that the principal symbol of the operator P under study is written as

$$p_m(x,t;\xi,\tau)=\prod_{j=1}^m (\tau-\lambda_j(x,t\,;\xi))\quad\text{with}\quad \lambda_j\in T(1).$$

We set $\partial_j = D_t - \lambda_j(x,t;D_x)$.

d) $\mathfrak{S}$ will denote the set of permutations $I=(i_1,\dots,i_k)$ of all the subsets of $(1,2,\dots,m)$; we write $|I|=k$ and $I\subset J=\{j_1,\dots,j_\ell\}$ if $|I|=0$ or if

$$\{i_\alpha : 1\leqslant\alpha\leqslant|I|\}\subset\{j_\beta : 1\leqslant\beta\leqslant|J|\}\ .$$

Now for $I = (i_1,\dots,i_k)$ we set

$$\partial_I = \begin{cases} \text{Identity if } |I| = 0 \\ \partial_{i_1} \circ \dots \circ \partial_{i_k} \quad \text{if } |I| \geq 1 . \end{cases}$$

e) If $A \in OPT(r)$, $B \in OPT(r')$ have symbols a and b, we denote by $A \otimes B$ the pseudo-differential operator whose symbol is $a.b$. We then set

$$P_I = \begin{cases} \text{Identity if } |I| = 0 \\ \partial_{i_1} \otimes \dots \otimes \partial_{i_k} \quad \text{if } I = (i_1,\dots,i_k), k \geq 1 . \end{cases}$$

Step 1 : Carleman estimates for ∂_j

The main result of this step is the following :

Proposition 1.4 :

There exist positive constants C, k_o, T_o, r such that for $k \geq k_o$, $T \leq T_o$, we have

(1.5) $$k|||v|||_o^2 \leq C|||\partial_j v|||_o^2 \ , \ 1 \leq j \leq m$$

and for the complex roots λ_j ,

(1.6) $$|||v|||_1^2 \leq C(1+kT^2)\ |||\partial_j v|||_o^2$$

for every $v \in C^\infty$ with $\text{supp}\, v \subset \{(x,t) : 0 \leq t \leq T, |x| \leq r\}$.

Proof :

Let us set $w = e^{\frac{1}{2}k(t-T)^2} . v$ and $\lambda_j(x,t\ ;\ D_x) = A + iB$ where $A,B \in OPT(1)$ have real symbols. Then

$$e^{\frac{1}{2}k(t-T)^2} \partial_j v = (D_t - A - iB + ik(t-T))w = Xw + Yw$$

with $X = D_t - A$, $Y = -iB + ik(t-T)$. We get

$$I = \int_0^T e^{k(t-T)^2} \|\partial_j v\|^2 dt = \int_0^T \|Xw\|^2 dt +$$

(1.7)
$$+ \int_0^T \|Yw\|^2 dt + 2\mathrm{Re} \int_0^T (Xw, Yw)\, dt$$

where $(\ ,\ )$ denotes the $L^2(\mathbb{R}^n)$ - inner product.
It is easy to see that

(1.8) $$2\mathrm{Re} \int_0^T (D_t w, ik(t-T)w)\, dt = k \int_0^T \|w\|^2\, dt .$$

On the other hand ,

$$- 2\mathrm{Re} \int_0^T (Aw,\ ik(t-T)w)\, dt = -2k\ \mathrm{Im} \int_0^T (t-T)(Aw, w)\, dt.$$

Now since $(Aw,w) = (w, A^*w)$ we have $\mathrm{Im}((A+A^*)w,w) = 0$.
Writing $A = \frac{1}{2}(A+A^*) + \frac{1}{2}(A-A^*)$ we get $\mathrm{Im}(Aw,w) = \frac{1}{2}\mathrm{Im}((A-A^*)w,w)$. The symbol of A being real, $A-A^*$ is of order zero so

(1.9) $$\left| 2\mathrm{Re} \int_0^T (Aw,\ ik(t-T)w)\, dt \right| \leq ckT \int_0^T \|w\|^2\, dt.$$

Now if λ_j is real we have $B \equiv 0$ so by (1.7), (1.8) and (1.9), taking $CT \leq \frac{1}{2}$ we get (1.5).
Let us look at the non-real roots λ_j ; it means that ∂_j is an elliptic p.d.o. We have with $L = A^*B$

$$① = 2\mathrm{Re} \int_0^T (Aw,\ iBw)\, dt = 2\mathrm{Im} \int_0^T (Aw, Bw)\, dt = 2\mathrm{Im} \int_0^T (w, Lw)\, dt.$$

The argument used before shows that $① = \mathrm{Im} \int_0^T (w,(L-L^*)w)\, dt$

so

(1.10) $$\left| 2\mathrm{Re} \int_0^T (Aw, iBw)\, dt \right| \leq C \int_0^T \|w\| \cdot \|w\|_1\, dt$$

where $\|\ \|_1$ is the $H^1(\mathbb{R}^n_x)$ norm.

Let us now check the last term

$$2\mathrm{Re}\int_0^T (D_t w, iBw)\,dt = -2\mathrm{Im}\int_0^T (D_t w, Bw)\,dt .$$

We have

$$\int_0^T (D_t w, Bw) = \int_0^T (w, [D_t, B]w)\,dt + \int_0^T (B^* w, D_t w)\,dt$$

so

$$\mathrm{Im}\int_0^T (D_t w, (B+B^*)w)\,dt = \mathrm{Im}\int_0^T (w, [D_t, B]w)\,dt = \mathrm{Re}\int_0^T (w, \frac{dB}{dt}w)\,dt .$$

We get

$$\mathrm{Im}\int_0^T (D_t w, Bw)\,dt = \frac{1}{2}\int_0^T (D_t w, (B-B^*)w)\,dt + \mathrm{Re}\int_0^T (w, \frac{dB}{dt}w)\,dt$$

and $B-B^*$ is in $T(0)$. We deduce from that

$$-2\mathrm{Im}\int_0^T (D_t w, Bw)\,dt \geq -C\int_0^T ||w|| \, (||D_t w|| + ||w||_1)\,dt.$$

On the other hand, $||D_t w|| \leq ||(D_t - A)w|| + ||w||_1$ and for all $\varepsilon > 0$ there exists $C_\varepsilon > 0$ such that

$$||w|| \, . \, ||(D_t - A)w|| \leq \varepsilon \, ||(D_t - A)w||^2 + C_\varepsilon ||w||^2 .$$

It follows that

$$(1.11) \quad -2\mathrm{Im}\int_0^T (D_t w, Bw)\,dt \geq -C_\varepsilon \int_0^T ||(D_t - A)w||^2\,dt -$$

$$- C_\varepsilon \int_0^T ||w||^2\,dt - C\int_0^T ||w|| \, . \, ||w||_1\,dt .$$

Taking ε small and k big enough, we deduce from (1.7)...

...(1.11) that

$$I \geq \frac{1}{2} \int_0^T \|Xw\|^2\, dt + \int_0^T \|Yw\|^2\, dt +$$

$$(1.12) \qquad + k \int_0^T \|w\|^2\, dt - C \int_0^T \|w\| \cdot \|w\|_1\, dt.$$

Now, the operator B being elliptic, we can find $E \in OPT(-1)$ such that $EB = Id + R_{-1}$ with $R_{-1} \in OPT(-1)$. Denoting by Λ the p.d.o with symbol $(1+|\xi|^2)^{1/2}$, it follows that

$$\|w\|_1 \leq \|EB\Lambda w\| + C\|w\| \leq \|E[B,\Lambda]w\| +$$

$$+ \|E\Lambda Bw\| + C\|w\| \leq C'(\|Bw\| + \|w\|)$$

so we get

$$(1.13) \qquad \|w\|_1 \leq C'(\|Bw - k(t-T)w\| + (1+kT)\|w\|).$$

For every $\varepsilon > 0$ we can then find $c_\varepsilon > 0$ such that

$$\int_0^T \|w\|_1 \cdot \|w\|\, dt \leq \varepsilon \int_0^T \|Bw-k(t-T)w\|^2\, dt +$$

$$+ c_\varepsilon(1+kT) \int_0^T \|w\|^2\, dt.$$

Recalling that $Y = -iB + ik(t-T)$, it follows from (1.12), taking ε small, that

$$I \geq c_1 \left(\int_0^T \|Xw\|^2\, dt + \int_0^T \|Yw\|^2 dt + k \int_0^T \|w\|^2\, dt \right) -$$

$$- c'(1+kT) \int_0^T \|w\|^2\, dt.$$

If T^{-1} and k are big enough we get

$$(1.14) \qquad I \geqslant c\left(\int_0^T (\|Xw\|^2 + \|Yw\|^2)dt + k \int_0^T \|w\|^2\, dt\right).$$

Now, from (1.13) and (1.14) we deduce

$$(1.15) \qquad \int_0^T \|w\|_1^2\, dt \leqslant$$

$$\leqslant c\left(\int_0^T \|Yw\|^2\, dt + (1+k^2T^2) \int_0^T \|v\|^2\, dt\right) \leqslant c'(1+kT^2)\, I.$$

Finally

$$\int_0^T \|D_t w\|^2\, dt \leqslant \int_0^T \|\partial_j w\|^2\, dt + \int_0^T \|w\|_1^2\, dt ,$$

so the proof of Proposition 1.4 is complete.

<u>Lemma 1.5</u>:

<u>There exists a positive constant</u> C <u>such that</u>

$$(1.16) \qquad (1+k^2T^2)\; |||v|||_0^2 \leqslant C|||D_t v|||_0^2$$

<u>for every</u> $v \in C^\infty(\mathbb{R}^{n+1})$ <u>with</u> $\operatorname{supp} v \subset \{(x,t) : 0 \leqslant t \leqslant \frac{T}{2}, |x| \leqslant r\}$.

<u>Proof</u> :

This inequality is equivalent to

$$(1 + k^2T^2) \int_0^T \|w\|^2\, dt \leqslant c \int_0^T \|D_t w + ik(t-T)w\|^2\, dt.$$

Now we see easily that

$$\int_0^T \|D_t w + ik(t-T)w\|^2 dt = \int_0^T \|D_t w\|^2 dt +$$

$$+ k^2 \int_0^T (t-T)^2\, \|w\|^2 dt + k \int_0^T \|w\|^2 dt ,$$

and if $t \leqslant \frac{T}{2}$, we have $(t-T)^2 \geqslant \frac{1}{4} T^2$.

Step 2 : Carleman estimate for $\partial_1 o \ldots o \partial_m$

Proposition 1.6 :

There exist positive constants C, k_o, T_o, r such that for $k \geqslant k_o$, $T \leqslant T_o$, we have

(1.17) $$|||v|||^2_{m-1} \leqslant C(\frac{1}{k} + T^2) \; |||\partial_1 o \ldots o \partial_m v|||^2_o$$

for every $v \in C^\infty(\mathbb{R}^{n+1})$, $\operatorname{supp} v \subset \{(x,t) : 0 \leqslant t \leqslant T/2, |x| \leqslant r\}$.

First of all, using the condition (H.4), we write

(1.18) $$\partial_1 o \ldots o \partial_m = P_1 \; o \; P_2 \; o \; P_3 + R, \quad d^o_= P_j = m_j$$

where P_1 is made of distinct elliptic factors, same thing for P_2, P_3 has only distinct hyperbolic factors (which means factors with real symbols) and R is such that

$$|||Rv|||^2 \leqslant C \; |||v|||^2_{m-1} \, .$$

This is possible using the following :

Lemma 1.7 :

Let I be in $\mathfrak{S}$ and $\tilde{I}$ be a permutation of I. Then there exists $a_j \in OPT(1)$ such that

(1.19) $$\partial_I - \partial_{\tilde{I}} = \Sigma \; a_J \; \partial_J$$

the sum being taken over all J such that $J \subset I$ and $|J| \leqslant |I| - 2$.

Proof :

a) We first prove the following :

Let A be in $OPT(1)$ and $J = \{j_1, \ldots, j_k\}$; then

$$\partial_{j_1} o \ldots o \partial_{j_p} \; o \; A \; o \; \partial_{j_{p+1}} \; o \ldots o \partial_{j_k} = A \; o \; \partial_1 o \ldots o \partial_{j_k} + \Sigma \, b_L \; \partial_L$$

the sum being taken over $L \subset J$, $|L| < |J|$.

To prove that, we use an induction on $|J|$ and we suppose it is true for $|J| \leq k$. We write

$$\partial_{j_1} o \ldots o \partial_{j_p} o A o \ldots o \partial_{j_{k+1}} = \partial_{j_1} o [\partial_{j_2} o \ldots o \partial_{j_p} o A o \ldots o \partial_{j_{k+1}}]$$

and we apply the induction to the operator inside the brackets.

b) To show (1.19) we also use an induction on $|I|$. Take $I=(i_1,\ldots,i_{k+1})$, $\tilde{I} = (j_1,\ldots,j_{k+1})$ a permutation of I and suppose $i_p = j_{k+1}$. We have

$$\partial_I = \partial_{i_1} o \ldots o \partial_{i_{p-1}} o (\partial_{j_{k+1}} o \ldots o \partial_{i_{k+1}})$$

or

$$\partial_I = (\partial_{j_{k+1}} o \partial_{i_2} o \ldots o \partial_{i_k}) o \partial_{i_{k+1}} .$$

In both cases we apply the induction to the bracket. In the first one get

$$\partial_I = \partial_{i_1} o \ldots o \partial_{i_{p-1}} o \partial_{i_{p+1}} o \ldots o \partial_{i_{k+1}} o \partial_{j_{k+1}} +$$

$$+ \partial_{i_1} o \ldots o \partial_{i_{p-1}} o \Sigma a_J \partial_J$$

where $J \subset \{i_p,\ldots,i_{k+1}\}$ and $|J| \leq k-p+1$. Using a) we get

$$\partial_I - \partial_{\tilde{I}} = (\partial_{i_1} o \ldots o \partial_{i_{p-1}} o \partial_{i_{p+1}} o \ldots o \partial_{i_{k+1}} - \partial_{j_1} o \ldots o \partial_{j_k}) o \partial_{j_{k+1}} +$$

$$+ \sum_{|J| \leq k} c_J \partial_J .$$

We then apply one more time the induction to the bracket to conclude. We use the same argument for the second case.
Let us now come back to the Carleman estimate. We first prove the following

$$(1.20) \quad (1+kT^2)^{m_1-|\alpha|-1} k^{m_1-|\alpha|} |||v|||^2_{|\alpha|} \leq C |||P_1 v|||^2_o , \ |\alpha| \leq m_1$$

for every $v \in C^{\infty}$, $\text{supp}\, v \subset \{0 \leq t \leq T/2,\ |x| \leq r\}$, where P_1 is given by (1.18).

a) We prove (1.20) for $|\alpha| = 0$ and every $m_1 \geq 1$.
We need first the following

Lemma 1.8 :

Suppose $\lambda_i(x,t;\xi) \neq \lambda_j(x,t;\xi)$ for all $(x,t;\xi) \in V_o \times \mathbb{R}^n \backslash 0$; then for every first order p.d.o, of the form $L = D_t + a(x,t;D_x)$ with $a \in OPT(1)$, there exist A,B,C in $OPT(0)$ such that

$$L = A \circ \partial_i + B \circ \partial_j + C . \tag{1.21}$$

Proof :

We just have to identify the symbols and use the hypothesis∎

Now we have, writing $P_1 = \partial_I$, for $1 \leq \ell \leq m_1 - 1$

$$(1+k^2T^2)^{\ell} |||v|||_o^2 \leq C \sum_{\substack{|J| \leq \ell \\ J \subset I}} |||\partial_J v|||_o^2 . \tag{1.22}$$

For $\ell = 1$ we use (1.16) and (1.21). Now suppose (1.22) is true up to the order $\ell - 1$ with $\ell \leq m_1 - 1$. Since $\ell - 1 \leq m_1 - 2$ then, for every $J \subset I$ with $|J| \leq \ell - 1$, we can find $\alpha, \beta \in I \backslash J$ such that $\{\alpha\} \cup \{\beta\} \cup J \subset I$. Using the induction, (1.16) and (1.21) we get

$$(1+k^2T^2)^{\ell} |||v|||_o^2 \leq C_1 (1+k^2T^2) \sum_{\substack{|J| \leq \ell - 1 \\ J \subset I}} |||\partial_J v|||_o^2$$

$$\leq C_2 \sum_J |||D_t \partial_J v|||_o^2$$

$$(1+k^2T^2)^{\ell} |||v|||_o^2 \leq C_2 \sum_J |||(A\partial_\alpha + B\partial_\beta + C)\partial_J v|||_o^2 \leq$$

$$\leq C \sum_{\substack{|K| \leq \ell \\ K \subset I}} |||\partial_K v|||_o^2$$

which proves (1.22) up to the order ℓ .
Now take $\ell = m_1 - 1$, choose for every J such that $|J| \leq m_1 - 1$ an index γ in $I \setminus J$ and apply (1.5) ; we get

$$(1.23) \quad k(1+k^2T^2)^{m_1-1} |||v|||_o^2 \leq C \sum_{\substack{|J| \leq m_1 \\ J \subset I}} |||\partial_J v|||_o^2 .$$

Now using Lemma 1.7 we get from (1.23)

$$(1.24) \quad k(1+k^2T^2)^{m_1-1} |||v|||_o^2 \leq$$

$$\leq C\{|||P_1 v|||_o^2 + \underbrace{\sum_{\substack{|J| \leq m_1-2 \\ J \subset I}} |||\partial_J v|||_1^2}_{①} \} .$$

Since $|J| \leq m_1 - 2$, using (1.21) with $L = D_x$ and D_t we get

$$(1.25) \quad \underbrace{\sum_{\substack{|J| \leq m_1-2 \\ J \subset I}} |||\partial_J v|||_1^2}_{①} \leq C \underbrace{\sum_{\substack{|K| \leq m_1-1 \\ K \subset I}} |||\partial_K v|||_o^2}_{②} .$$

Using (1.5) with $j \in I \setminus K$, since $|K| \leq m_1 - 1$ we get

$$② \leq \frac{c}{k} \sum_{\substack{|L| \leq m_1 \\ L \subset I}} |||\partial_L v|||_o^2 .$$

It follows from Lemma 1.7 that

$$② \leq \frac{c}{k} \{|||P_1 v|||_o^2 + \sum_{\substack{|J| \leq m_1-1 \\ J \subset I}} |||\partial_J v|||_o^2 + \sum_{\substack{|J| \leq m_1-2 \\ J \subset I}} |||\partial_J v|||_1^2\}$$

which can be written

$$② \leq \frac{c}{k} \{|||P_1 v|||_o^2 + ② + ①\} .$$

Using (1.25) we deduce, taking k big enough ,

$$① \leq \frac{c}{k} |||P_1 v|||_o^2 .$$

It follows from (1.24) that

$$k(1+k^2T^2)^{m_1-1} |||v|||_o^2 \leqslant C|||P_1 v|||_o^2$$

which proves a).

b) Inequality (1.20) is true for $m_1 = 1$.
Indeed, in this case (1.20) follows from (1.5) and (1.6).

c) We prove now that (1.20) for m_1 and $0 \leqslant |\alpha| \leqslant m_1$ implies (1.20) for m_1+1 and $1 \leqslant |\alpha| \leqslant m_1+1$.
Let us set $P_{m_1+1} = \partial_I$ with $|I| = m_1+1 \geqslant 2$. With $i,j \in I$, $i \neq j$, we set $I_1 = I\backslash\{i\}$, $I_2 = I\backslash\{j\}$. Using the induction we get with $a,b,c \in OPT(0)$:

$$(1+kT^2)^{m_1-|\alpha|-1} k^{m_1-|\alpha|} |||D^\alpha(b\partial_j)v|||_o^2 \leqslant C|||\partial_{I_2}(b\partial_j)v|||_o^2$$

with $0 \leqslant |\alpha| \leqslant m_1$. Writing, by means of Lemma 1.8, $D = a\partial_i + b\partial_j + c$ we get with $1 \leqslant |\beta| = |\alpha| + 1 \leqslant m_1+1$

$$(1+kT^2)^{m_1-|\beta|} k^{m_1+1-|\beta|} |||D^\beta v|||_o^2 \leqslant \sum_{\substack{J \subset I \\ |J| \leqslant |I|}} |||\partial_J v|||_o^2 .$$

Using Lemma 1.7 we deduce that

$$(1+kT^2)^{m_1-|\beta|} k^{m_1+1-|\beta|} |||D^\beta v|||_o^2 \leqslant$$

$$\leqslant C(|||P_{m+1}v|||_o^2 + \sum_{|J| \leqslant m_1-1} |||\partial_J v|||_1^2) .$$

We then use the same argument as we used for (1.24) to prove our claim. With the notation of (1.18) we prove now

$$(1.26) \qquad (1+kT^2)^{m_1+m_2-|\alpha|-2} k^{m_1+m_2-|\alpha|} |||v|||_{|\alpha|} \leqslant$$

$$\leqslant C\ |||P_1\ P_2 v|||_o^2 ,\ 0 \leqslant |\alpha| \leqslant m_1+m_2 .$$

Indeed, using (1.20) with $P_2 v$ instead of v, we get

$$(1+kT^2)^{m_1-|\alpha|-1}\, k^{m_1-|\alpha|}\, |||D^\alpha P_2 v|||_o^2 \leqslant$$

$$\leqslant C\, |||P_1P_2v|||_o^2, \quad 0 \leqslant |\alpha| \leqslant m_1.$$

Now $[D^\alpha, P_2]$ is a differential operator in t, pseudo-differential in x, of order $m_2+|\alpha|-1$. It follows that

$$(1+kT^2)^{m_1-|\alpha|-1}\, k^{m_1-|\alpha|}\, |||P_2D^\alpha v|||_o^2 \leqslant$$

$$\leqslant C\,(|||P_1P_2v|||_o^2 + (1+kT^2)^{m_1-|\alpha|-1}\, k^{m_1-|\alpha|} \sum_{|\beta|<m_2+|\alpha|-1} |||D^\beta v|||_o^2).$$

Using (1.20) for P_2 instead of P_1 we get

$$(1+kT^2)^{m_2-|\gamma|-1}\, k^{m_2-|\gamma|}\, |||D^\gamma v|||_o^2 \leqslant C\, |||P_2 v|||_o^2, \quad 0 \leqslant |\gamma| \leqslant m_2.$$

Applying this inequality to $D^\alpha v$, $0\leqslant|\alpha|\leqslant m_1$ we get

$$(1+kT^2)^{m_1+m_2-|\alpha|-|\gamma|-2}\, k^{m_1+m_2-|\alpha|-|\gamma|}\, |||D^\gamma D^\alpha v|||_o^2 \leqslant$$

$$\leqslant C(|||P_1P_2v|||_o^2 + (1+kT^2)^{m_1-|\alpha|-1}\, k^{m_1-|\alpha|} \sum_{|\beta|<m+|\alpha|-1} |||D^\beta v|||_o^2).$$

Now the coefficient, in the left hand side, of $|||D^\beta v|||_o^2$ for $|\beta| < m_2 + |\alpha|-1$, is equal to $(1+kT^2)^{m_1+m_2-|\beta|-2}\, k^{m_1+m_2-|\beta|}$.

But we have

$$m_1+m_2-|\beta|-2 \geqslant m_1+|\alpha| \qquad \text{and} \quad m_1+m_2-|\beta| \geqslant m_1-|\alpha|+2,$$

so taking k big enough we deduce, from that, (1.26). Using (1.26) we are able to prove

$$(1.27) \qquad k(1+kT^2)^{-1}\, |||v|||_{m_1+m_2+m_3-1}^2 \leqslant C\, |||P_1P_2P_3v|||_o^2 .$$

Let us remark that Proposition 1.6 follows from (1.27) and (1.18). To prove this inequality we use an induction on m_3. For $m_3 = 0$ (1.27) reduces to (1.26). The case $m_3 = 1$ will be proved by an induction on $m_1 = d^{\underline{o}} P_1$. For $m_1 = 0$ we use (1.20) with P_2 and $P_3 v = \partial v$ instead of v. It follows that

$$(1+kT^2)^{m_2-|\alpha|-1} k^{m_2-|\alpha|} |||\partial v|||^2_{|\alpha|} \leq C|||P_2\partial v|||^2_o \; ; |\alpha| \leq m_2 .$$

We use then (1.5) to get

$$(1+kT^2)^{m_2-|\alpha|-1} k^{m_2-|\alpha|+1} |||v|||^2_{|\alpha|} \leq C\, |||P_2\partial v|||^2_o .$$

Taking $|\alpha| = m_2$ we get the case $m_1 = 0$. Let us suppose (1.27) true (when $m_3 = 1$) for $d^{\underline{o}} P_1 \leq m_1$ and take $d^{\underline{o}} P_1 = m_1 + 1 \geq 2$. Using (1.26) with $P_1 P_2$, $|\alpha| = m_1 + m_2$, $P_3 v = \partial v$ instead of v, we can write

$$(1.28) \qquad (1+kT^2)^{-1}\, k\, |||\partial v|||^2_{m_1+m_2} \leq C\, |||P_1P_2\partial v|||^2_o .$$

Now let ∂_1 be a factor of P_1 and $\hat{P}_1 = P_1 \setminus \partial_1$; then we can apply the induction to $\hat{P}_1$ (which is of $d^{\underline{o}}\, m_1$) and $\partial_1 v$ instead of v. It follows, with $P_3 = \partial$,

$$(1+kT^2)^{-1}\, k\, |||\partial_1 v|||^2_{m_1+m_2} \leq C|||\hat{P}_1 P_2 P_3 \partial_1 v|||^2_o .$$

Using Lemma 1.7 it follows easily that

$$(1.29) \quad (1+kT^2)^{-1}\, k\, |||\partial_1 v|||^2_{m_1+m_2} \leq C|||P_1P_2\partial v|||^2_o .$$

Using (1.28), (1.29) and Lemma 1.8 we get

$$(1+kT^2)\, k|||v|||^2_{m_1+m_2+1} \leq C\, |||P_1P_2\partial v|||^2_o .$$

The case $m_3=1$ is thus completely proved. Suppose (1.27) true for $d^o P_3 \leq m_3$ and suppose $d^o P_3 = m_3+1 \geq 2$. Let ∂_1 and ∂_2 be two distinct factors of P_3 and set $\hat{P}_3 = P_3 \quad \partial_1$, $\tilde{P}_3 = P_3 \quad \partial_2$ which are of degree m_3. We use the induction with $\hat{P}_3$ (resp. $\tilde{P}_3$) and $\partial_1 v$ (resp. $\partial_2 v$) instead of v ; it follows that

$$(1+kT^2)^{-1} k \, |||\partial_1 v|||^2_{m_1+m_2+m_3-1} \leq C \, |||P_1 P_2 \hat{P}_3 \, \partial_1 v|||^2_o$$

and the same for $\partial_2 v$. Then, using Lemma 1.7, Lemma 1.8 and taking $(\frac{1}{k} + T^2)$ small enough, we get (1.27) when $d^o P_3 = m_3+1$.

Q.E.D

<u>Step 3</u> : <u>Expression of</u> $\tilde{P}_m - \partial_1 o \ldots o \partial_m$

This section is devoted to the proof of the following

<u>Proposition 1.9</u> :

<u>There exist</u> p.d.o. $a_J \in OPT([\frac{m-|J|}{2}])$ <u>such that</u>

(1.30) $$P_m(x,t;D_x,D_t) = \partial_1 o \ldots o \partial_m + \sum_{|J| \leq m-1} a_J \, \partial_J$$

<u>where</u> [a] <u>denotes the entire part of</u> a.

This result will be a consequence of the following :

<u>Proposition 1.10</u> :

<u>Let</u> Q <u>be in</u> T(r) <u>and</u> $I \in \mathfrak{S}$. <u>We can find</u>

p.d.o. $Q_{I,J} \in OPT(r + [\frac{|I| - |J|}{2}])$ <u>such that</u>

(1.31) $$Q \otimes P_I = Q \, o \, \partial_I + \sum_{\substack{|J| \leq |I| \\ J \subset I}} Q_{I,J} \, o \, \partial_J .$$

Proof :

Let us recall that P_I denotes the p.d.o. with symbol $\prod_{j \in I}(\tau-\lambda_j)$.

a) For every $R \in OPT(r)$ and $I \subset \{1,2,\ldots,m\}$ we can find symbols r_{IJ} belonging to $T(r+[\frac{|I|-|J|}{2}])$ such that

$$(1.32) \qquad \sigma(R \circ \partial_I) = \sigma(R).\sigma(P_I) + \sum_{\substack{|J|<|I| \\ J \subset I}} r_{IJ}\, \sigma(P_J).$$

Indeed, it is obvious for $|I|=1$; setting $\partial_I = \partial \circ \partial_{I_o}$

we get

$$R \circ \partial_I = R \circ \partial \circ \partial_{I_o} = [R,\partial] \circ \partial_{I_o} + \partial \circ R \circ \partial_{I_o}.$$

Using the induction we get

$$\sigma([R,\partial] \circ \partial_{I_o}) = a.\sigma(P_{I_o}) + \sum_{\substack{|J|<|I_o| \\ J \subset I_o}} a_{I_oJ}\, \sigma(P_J)$$

where $a \in T(r)$ and $a_{I_oJ} \in T(r + \left[\frac{|I_o|-|J|}{2}\right]) \subset T(r+\left[\frac{|I|-|J|}{2}\right])$.

On the other hand,

$$\sigma(\partial \circ R \circ \partial_{I_o}) = \sigma(\partial).\sigma(R \circ \partial_{I_o}) +$$

$$\sum_{|\alpha|+k\neq 0} \frac{1}{\alpha!k!}\, \partial_\xi^\alpha \partial_\tau^k \sigma(\partial)\, D_x^\alpha D_t^k\, \sigma(R \circ \partial_{I_o}) = ① + ② .$$

The induction implies

$$① = \sigma(\partial)\,\sigma(R)\,\sigma(P_{I_o}) + \sigma(\partial) \sum_{|J|<|I_o|} r_{I_oJ}\,\sigma(P_J).$$

The first term in the right hand side is precisely $\sigma(R)\,\sigma(P_I)$. In the second one, we have $\sigma(\partial)\,\sigma(P_J) = \sigma(P_{J'})$ with $|J'| = |J| + 1$ and

$$r_{I_oJ} \in T\left(r + \left[\frac{|I_o| - |J|}{2}\right]\right) = T\left(r + \left[\frac{|I| - |J'|}{2}\right]\right).$$

Now, since $\sigma(\partial) = \tau - \lambda$ with $\lambda \in T(1)$, we can write

$$② = c_o \frac{\partial}{\partial t}\,\sigma(R \circ \partial_{I_o}) + \sum_{\alpha\neq 0} c_\alpha\, \partial_\xi^\alpha\, \lambda\, D_x^\alpha\, \sigma(R \circ \partial_{I_o}) = ③ + ④.$$

Using (1.32) with I_o we get

$$③ = a.\,\sigma(P_{I_o}) + \sum_{|J|=|I_o|-1} b_J\,\sigma(P_J) +$$

$$+ \sum_{|J|<|I_o|} c_{JI_o}\,\sigma(P_J) + \sum_{|J|\leqslant|I_o|-1} d_{I_oJ} \sum_{|L|=|J|-1} \sigma(P_L)$$

with $a \in T(r)$, $b_J \in T(r+1)$, $c_{JI_o} \in T\left(\left[\frac{|I_o| - |J|}{2}\right] + r\right)$,

$$d_{I_oJ} \in T\left(r+1 - \left[\frac{|I_o| - |J|}{2}\right]\right).$$

It follows that

$$③ = \sum_{|J|<|I|} r_{IJ}\,\sigma(P_J)\ ,\ r_{IJ} \in T\left(r + \left[\frac{|I|-|J|}{2}\right]\right).$$

To estimate ④ we also use (1.32) ; it follows that

$$④ = \sum_{\alpha \neq 0} q_{1-|\alpha|}\, D_x^{\alpha}(\sigma(R) .\ \sigma(P_{I_o})) +$$

$$+ \sum_{\alpha \neq 0} q_{1-|\alpha|}\, D_x^{\alpha}\Big(\sum_{|J|<|I_o|} r_{I_oJ}\,\sigma(P_J)\Big) = ⑤ + ⑥ .$$

Using Leibniz formula we get

$$⑤ = \sum_{k=1}^{|I_o|} \sum_{|J|=|I_o|-k} \Big(\sum_{|\alpha|\geq k} q_{\alpha Jk}\Big)\, \sigma(P_J)$$

with $q_{\alpha Jk} \in T(r+1-|\alpha|+k)$. There exists $q_{kJ} \in T(r+1)$ such that

$$q_{kJ} \sim \sum_{|\alpha|\geq k} q_{\alpha Jk} .$$

It follows that

$$⑤ = \sum_{|J|<|I_o|} q_{I_oJ}\, \sigma(P_J)$$

with $q_{I_oJ} \in T(r+1) \subset T\left(r + \left[\frac{|I|-|J|}{2}\right]\right)$, since $|I|-|J| \geq 2$.

We use the same technique for ⑥. Here the coefficient of $\sigma(P_L)$ for $|L| \leq |I_o|-1$ belongs to $T\left(r+\left[\frac{|I_o|-|L|+1}{2}\right]\right)$.

This proves (1.32).

b) We now prove

$$(1.33) \qquad Q \circ \partial_I = Q \otimes P_I - \sum_{\substack{|J| < |I| \\ J \subset I}} Q_{IJ} \circ \partial_J$$

$Q_{IJ} \in OPT(r+\left[\frac{|I|-|J|}{2}\right])$.

It is obviously true for $|I| = 1$, so assume it is true up to $|I|-1$. By (1.32)

$$\sigma(Q \circ \partial_I) = \sigma(Q)\,\sigma(P_I) + \sum_{|J|<|I|} \sigma(Q_{IJ} \otimes P_J).$$

By the induction

$$\sigma(Q \circ \partial_I) = \sigma(Q)\,\sigma(P_I) + \sum_{|J|<|I|} \{\sigma(Q_{IJ} \circ \partial_J) +$$

$$+ \sum_{|K|<|J|} \sigma(Q_{IJK} \circ \partial_K)\}$$

where

$$Q_{IJ} \in OPT(r+\left[\frac{|I|-|J|}{2}\right]) \quad \text{and}$$

$$Q_{IJK} \in OPT(r+\left[\frac{|I|-|J|}{2}\right] + \left[\frac{|J|-|K|}{2}\right]) \subset OPT(r+\left[\frac{|I|-|K|}{2}\right]),$$

so (1.33) is true up to the order $|I|$.

Q.E.D.

Now, theorem 1.3 follows easily from Proposition 1.6 and Proposition 1.9∎

§2. ELLIPTIC OPERATORS WITH NON-SMOOTH DOUBLE CHARACTERISTICS.

One of the conditions used to prove Theorem 1.1 was the smoothness of the characteristic roots. It is a natural question to ask if this condition can be relaxed. The following example shows that we must, in general, answer by the negative. Consider in $\mathbb{R}^3$ the differential operator

$$(2.1) \qquad P = (D_t^2 + D_x^2 + D_y^2)^2 - \frac{1}{4} D_x^4 + t(D_x^2 + D_y^2)^2 .$$

It is elliptic for $|t|$ small and it has an, at most double, characteristic root, but it can be shown that there exists a zero order perturbation $a \in C^\infty$ near the origin such that $P + a$ does not possess the uniqueness property with respect the surface $t = 0$. However in some cases the smoothness of the roots can be relaxed, as shows the following :

Theorem 2.1:

Let P be a second order differential operator in a neighborhood V of a point $x_0 \in \mathbb{R}^2$ with analytic principal part and bounded lower order terms. Let $S = \{x \in V : \Phi(x) = \Phi(x_0)\}$, a C^2 non-characteristic hypersurface near x_0. Then there exists a neighborhood W of the point x_0 such that every $u \in C^\infty(V)$ satisfying

$$\begin{cases} Pu = 0 \quad \text{in} \quad V \\ u = 0 \ \text{in} \quad \{x \in V : \Phi(x) \leq \Phi(x_0)\} \end{cases}$$

vanishes in W.

This result will follow from a slightly stronger one, namely

Proposition 2.2 :

In a neighborhood V of the origin in $\mathbb{R}^2_{(x,t)}$

let us consider the elliptic operator

$$P = D_t^2 + 2b\, D_x\, D_t + cD_x^2 + \alpha\, D_x + \beta\, D_t + \gamma$$

with $b, c \in C^\infty$ and $\alpha, \beta, \gamma \in L^\infty$ and set $\Delta = b^2 - c$. Let us suppose that the function $t \to \Delta(0,t)$ has, at $t=0$ a zero of finite order. Then there exists a neighborhood W of the origin in which every $u \in C^\infty(V)$ such that $Pu=0$ and $u_{|t<0}=0$ vanishes.

Indeed, setting $x'=x_1$, $t=\langle x, \text{grad}\ \Phi(x_0)\rangle + A\,|x|^2$ where A is a big enough positive constant, Theorem 2.1 reduces to Proposition 2.2.

Proof of proposition 2.2

Step 1 : It is sufficient to prove Proposition 2.2 assuming that

(2.2) $$\Delta(x,t) = t^\ell\, \Delta_1(x,t),\ \ell \in \mathbb{N},\ \Delta_1(0,0) \neq 0.$$

We use the same argument as in Chap 1 theorem 4.1 slightly modified to take account of the fact that Δ may take complex values. Let us suppose Proposition 2.2 true under (2.2). By the Malgrange-Weierstrass theorem we can write for $|x| < r$ and $0 < t < T$

$$\Delta(x,t) = a(x,t)\ \{t^k + \sum_{j=0}^{k} a_j(x)\ t^{k-j}\}$$

with $a(0,0) \neq 0$, $a_j(0) = 0$.

We know that there exist k open sets $\mathcal{O}_1,\dots,\mathcal{O}_k$, included in $]-r,r[$, whose union is dense, such that in each $\mathcal{O}_j$, $\frac{\Delta}{a}$ has exactly j distinct roots which can be represented by j C^∞ functions $\rho_1(x),\dots,\rho_j(x)$. Let u be a solution of $Pu = 0$ such that $u_{|t<0}=0$.

Suppose $\operatorname{supp} u \cap]-r,r[\times]0,T[\neq \phi$; then there exists j and a connected component V of the $\mathcal{O}_j$ such that $\widehat{\operatorname{supp} \overset{o}{u}} \cap V\times]0,T[\neq \phi$. Shrinking V we can assume that :

(2.3)"For every open set $W \subset V$, $\operatorname{supp} u \cap W \times]0,T[\neq \phi$."

Let us write in V

$$\frac{\Delta}{a}(x,t) = \prod_{\ell=1}^{p} (t-\rho_\ell(x))^{\alpha_\ell} \prod_{\ell=p+1}^{j} (t-\rho_\ell(x))^{\alpha_\ell}$$

where $\rho_1,\ldots,\rho_p$ are the non-real roots and $\rho_{p+1},\ldots,\rho_j$ the real roots. Let $x_o \in V$ be such that $\operatorname{Im} \rho_1(x_o) \neq 0$. There exists $x_o \in V_1 \subset V$ such that $\operatorname{Im} \rho_1(x) \neq 0$ for $x \in V_1$; now in $V_1\times]0,T[$, $t - \rho_1(x) \neq 0$. We look at the second root ρ_2 in V_1. If $\operatorname{Im} \rho_2 \equiv 0$ then ρ_2 is real in V_1; if not, we shrink V_1 to V_2 where $\operatorname{Im} \rho_2 \neq 0$ so that $t-\rho_2(x) \neq 0$ in V_2 etc... up to ρ_p. We are now in the following situation : there exists an open set $\tilde{V} \subset V$ such that for $(x,t) \in \tilde{V}\times]0,T[$

$$\Delta(x,t) = b(x,t) \cdot \prod_{\ell\in J} (t-\rho_\ell(x))^{\alpha_\ell}$$

where $J \subset \{1,2,\ldots,j\}$, the roots ρ_ℓ are real and b does not vanish. For simplicity we will take $J = \{1,2,\ldots,q\}$ and we will suppose that in $\tilde{V}$ we have $\rho_1(x) <\ldots< \rho_q(x)$. Note that by (2.3), $\operatorname{supp} u \cap \tilde{V}\times]0,T[\neq \phi$. For $x \in \tilde{V}$ we define

$$\tilde{\rho}_o(x)=0,\ \tilde{\rho}_\ell(x) = \sup(0, \inf(T,\rho_\ell(x))),\ 1\leqslant\ell\leqslant q,\ \tilde{\rho}_{q+1}(x)=T$$

and

$$A_\ell = \{(x,t) \in \tilde{V}\times]0,T[\ : \ \tilde{\rho}_\ell(x) \leqslant t \leqslant \tilde{\rho}_{\ell+1}(x)\},\ 0 \leqslant \ell \leqslant q.$$

Let $\ell_o = \text{Min}\,\{\ell : \text{supp}\,u \cap A_\ell \neq \emptyset\}$.

a) Suppose there exists $(x_o,t_o) \in A_{\ell_o} \cap \text{supp}\,u$ with $t_o = \rho_{\ell_o}(x_o)$.

We perform the following change of coordinates : $x' = x - x_o$, $t' = t - \rho_{\ell_o}(x)$. In these coordinates, near the origin $\Delta = t'^k \tilde{\Delta}$ with $\tilde{\Delta}(0,0) \neq 0$, and the new function $\tilde{u}$ vanishes for $t' < 0$. By hypothesis $u=0$ near (x_o,t_o), which is a contradiction .

b) Suppose $\text{supp}\,u \cap A_{\ell_o} \cap \{t = \rho_{\ell_o}(x)\} = \phi$. We perform the same change of coordinates as before with x_o the mid-point of $\tilde{V}$. The set $\{(x,t) \in V \times]0,T[\;:\; t = \rho_{\ell_o}(x)\}$ is transformed into $\{(x,t),\ t'=0,\ |x'| < r\}$. Let

$\varepsilon_o = \text{Min}\,\{\varepsilon : \{\text{Graph of } t' - \varepsilon = \frac{-\varepsilon}{r^2} x'^2\} \cap \tilde{A}_{\ell_o} \cap \text{supp}\,\tilde{u} \neq \emptyset\}$.

By hypothesis $\varepsilon_o > 0$. We take $(x'_o, t'_o) \in A_{\ell_o} \cap \text{supp}\,\tilde{u}$ such that $t'_o = -\frac{\varepsilon_o}{r^2} x'^2_o + \varepsilon_o$. Near this point the operator P has simple roots since Δ is different from zero and the support of $\tilde{u}$ is on one side of the parabola $t' = -\frac{\varepsilon_o}{r^2} x'^2 + \varepsilon_o$. We conclude that $\tilde{u}$ vanishes near (x'_o, t'_o), which is a contradiction.

Step 2 : theorem 2.1 is true under the assumption (2.2).

Since the result is well known for $\ell = 0$, let us suppose that $\ell \geq 1$.

Let $P = \partial_t^2 + 2b\,\partial_x\,\partial_t + c\,\partial_x^2 + \alpha\,\partial_x + \beta\,\partial_t + \gamma$. We make near the origin the singular change of coordinates

$$X = x, t = (\delta - X^2)T, \ \delta > 0 \text{ small.}$$

It is easy to see that P transforms to an elliptic operator $\tilde{P}$ such that

$$(\delta-X^2)\tilde{P} = A\,\partial_T^2 + 2(\delta-X^2)B\,\partial_X\,\partial_T + C(\delta-X^2)^2\partial_X^2 + f\,\partial_T + g(\delta-X^2)\cdot\partial_X + h$$

where $f,g,h \in L^\infty$ and A,B,C are C^∞ functions of (X,T) such that

$$(2.4)\qquad \begin{cases} A \text{ is a real, } A(0,0)=1,\ \operatorname{Im} B(0,0)\neq 0 \\ (B^2-AC)(X,T)=(\delta-X^2)^{\ell+2}\,T^{\ell}\,D(X,T),\ D(0,0)\neq 0. \end{cases}$$

Writing for simplicity (x,t) instead of (X,T), it follows that we have

$$(2.5)\quad (\delta-x^2)\tilde{P} = A.P_1P_2 + \alpha\frac{\partial}{\partial t} + (\delta-x^2)(\beta+\gamma t^{\frac{\ell}{2}-1})\frac{\partial}{\partial x} + \lambda$$

where $\alpha,\beta,\gamma,\lambda \in L^\infty$ and

$$P_j = \frac{\partial}{\partial t} + (\delta-x^2)b\frac{\partial}{\partial x} + (-1)^j(\delta-x^2)^{\frac{\ell}{2}+1}t^{\frac{\ell}{2}}c\frac{\partial}{\partial x}\qquad j=1,2.$$

Here $b = \frac{B}{A}$ so that $\operatorname{Im} b(0,0)\neq 0$ and $c = \frac{D^{1/2}}{A}$. We prove now a Carleman estimate for P_j.

<u>Proposition 2.3</u> :

<u>There exists positive constants</u> C,γ_o,T_o,r <u>such that for</u> $\gamma\geqslant\gamma_o$ <u>and every</u> $u\in C^\infty$ <u>near the origin such that</u> $\operatorname{supp} u\subset\{(x,t) : 0\leqslant t\leqslant T_o,\ |x|\leqslant r\}$ <u>we have</u>

$$(2.6)\quad \gamma\|t^{-\gamma-1}u\|^2 + \frac{1}{\gamma}\|t^{-\gamma}(\delta-x^2)\partial_x u\|^2 + \frac{1}{\gamma}\|t^{-\gamma}\partial_t u\|^2 \leqslant C\,\|t^{-\gamma}P_j u\|^2$$

<u>where</u> $\|\ \|^2$ <u>is the</u> L^2 <u>norm.</u>

Proof :

Let us set $u=t^{\gamma}v$. It follows that

$$t^{-\gamma} P_j u = P_j v + \gamma t^{-1} v = Xv + Yv$$

where

$$(2.7)\quad \begin{cases} X = \partial_t + (\delta-x^2) b_1 \partial_x \ ; \ b_1 = \mathrm{Re}\, b + (-1)^j (\delta-x^2)^{\frac{\ell}{2}} t^{\frac{\ell}{2}} \mathrm{Re}\, c \\ Y = \gamma t^{-1} + i(\delta-x^2) b_2 \partial_x \ ; \ b_2 = \mathrm{Im}\, b + (-1)^j (\delta-x^2)^{\frac{\ell}{2}} t^{\frac{\ell}{2}} \mathrm{Im}\, c. \end{cases}$$

Now

$$(2.8)\quad \| t^{-\gamma} P_j u\|^2 = \|Xv\|^2 + \|Yv\|^2 + 2\mathrm{Re}\,(Xv,Yv).$$

It is easy to see that

$$(2.9)\qquad 2\mathrm{Re}(\partial_t v, \gamma t^{-1} v) = \gamma \| t^{-1} v\|^2$$

$$(2.10)\qquad |2\mathrm{Re}((\delta-x^2) b_1 \partial_x v, \gamma t^{-1} v)| \leq C\gamma \| t^{-\frac{1}{2}} v\|^2$$

$$(2.11)\qquad 2\mathrm{Re}((\delta-x^2) b_1 \partial_x v,\ i(\delta-x^2) b_2 \partial_x v) = 0 .$$

On the other hand,

$$(2.12)\quad 2\mathrm{Re}(\partial_t v, i(\delta-x^2) b_2 \partial_x v) = \underbrace{-(v, i(\delta-x^2) \frac{\partial b_2}{\partial t} \partial_x v)}_{1/} + \underbrace{(v,\ i\ \partial_x[(\delta-x^2) b_2] \partial_t v)}_{2/} .$$

Since $b_2(0,0) \neq 0$ we can write :

$$1/ = -\left(\frac{\partial_t b_2}{b_2} v,\ i(\delta-x^2) b_2 \partial_x v\right) = -\left(\frac{\partial_t b_2}{b_2} v, Yv\right) + \gamma\left(\frac{\partial_t b_2}{b_2} v,\ t^{-1} v\right).$$

It follows that for every $\varepsilon > 0$ we can find C_ε such that

$$(2.13)\ |1/| \leq \varepsilon \|Yv\|^2 + C_\varepsilon \|t^{-1} v\|^2 + C \cdot \gamma(\|t^{-\frac{1}{2}} v\|^2 + \|t^{\frac{\ell}{4}-1} v\|^2).$$

By (2.7) we have

$$2/ = (v, i\partial_x[(\delta - x^2)b_2] Xv) - (v, i\,\partial_x[(\delta - x^2)b_2](\delta - x^2)b_1\partial_x v) = 3/ + 4/.$$

Since $a = \partial_x[(\delta - x^2)b_2] \in L^\infty$ we get

$$|3/| \leq \varepsilon \|X v\|^2 + C_\varepsilon \|v\|^2 .$$

Now

$$4/ = -\left(\frac{ab_1}{b_2} v,\ i(\delta - x^2) b_2 \partial_x v\right)$$

and by the same argument which we used for 1/, we get

$$|4/| \leq \varepsilon \|Yv\|^2 + C_\varepsilon \|t^{-\frac{1}{2}} v\|^2.$$

It follows that

$$(2.14)\ |2/| \leq \varepsilon(\|Xv\|^2 + \|Yv\|^2) + C_\varepsilon \|t^{-1/2} v\|^2$$

Using (2.8)...(2.14) it is easy to see that for k and T_0^{-1} big enough

$$(2.15)\ \gamma \|t^{-1} v\|^2 + \|Xv\|^2 + \|Yv\|^2 \leq C \|t^{-\gamma} P_j u\|^2.$$

Now by (2.7)

(2.16) $\|\partial_t v\|^2 + \|(\delta-x^2)\partial_x v\| \leqslant \|Xv\|^2 + \|Yv\|^2 + \gamma^2\|t^{-1}v\|^2.$

From (2.15) and (2.16) we deduce (2.6).

Proposition 2.4 :

There exist positive constants C, γ_o, T_o, r such that for $\gamma \geqslant \gamma_o$ and $u \in C^\infty$ near the origin such that $\operatorname{supp} u \subset \{x,t) : 0 \leqslant t \leqslant T_o, |x| \leqslant r\}$

we have

(2.17) $$\gamma^2\|t^{-\gamma-2}u\|^2 \leqslant C\,\|(\delta-x^2)t^{-\gamma}\tilde{P}u\|^2.$$

Before we give the proof of Proposition 2.4, let us remark that (2.17) implies by a classical argument uniqueness of the Cauchy problem for $(\delta-x^2)\tilde{P}$ and then for the original operator P.

Proof of proposition 2.4 :

Since u is flat on $t=0$ it follows that P_2u is C^∞ and flat on $t=0$. By Proposition 2.3, applied twice, we can write

(2.18) $$\gamma^2\|t^{-\gamma-2}u\|^2 + \|t^{-\gamma-1}(\delta-x^2)\partial_x u\|^2 + \|t^{-\gamma-1}\partial_t u\|^2 \leqslant$$

$$\leqslant C\,\gamma\|t^{-\gamma-1}P_2u\|^2 \leqslant C'\|t^{-\gamma}P_1P_2u\|^2.$$

By (2.5)

(2.19) $$\|t^{-\gamma}P_1P_2u\|^2 \leqslant C\{\|t^{-\gamma}(\delta-x^2)\tilde{P}u\|^2 + \|t^{-\gamma}\partial_t u\|^2$$

$$+ \|(\delta-x^2)t^{-\gamma}\partial_x u\|^2 + \|t^{-\gamma-1+\frac{\ell}{2}}(\delta-x^2)\partial_x u\|^2 +$$

$$+ \|t^{-\gamma}u\|^2\}.$$

The inequality (2.17) then follows from (2.18) and (2.19) taking γ and T_o^{-1} big enough.

§3. THE TRIPLE COMPLEX CHARACTERISTIC CASE

In the two preceeding sections, the complex roots of the characteristic equation where at most double, while the real ones were simple (the case of real multiple characteristics will be discussed in the following chapter). We now discuss the triple complex roots and ask if the uniqueness holds without any condition on the lower order terms. The answer is then negative. Indeed, there exists an operator of the form

$$P = \Delta^3 + \sum_{|\alpha|\leq 5} a_\alpha(x)\, D_x^\alpha \tag{3.1}$$

where Δ is the Laplacian in $\mathbb{R}^n$ and a_α are *continuous* functions, which fails to have the uniqueness property. However, as we are going to see, it is not the size of the characteristics which is the cause of the non-uniqueness, but the regularity of the coefficients.

Theorem 3.1:

Let P be an $m^{\underline{th}}$ order differential operator in a neighborhood V of $x_o \in \mathbb{R}^n$ and $S = \{x \in V : \phi(x) = \phi(x_o)\}$ be a non-characteristic hypersurface. Let us suppose that

$$p_m(x,\xi+\tau N) = p_m(x,N) \prod_{j=1}^{\ell} (\tau-\lambda_j(x,\xi,N))^{\alpha_j}, \quad \alpha_j \in \mathbb{N} \tag{3.2}$$

for $(x,\xi,N) \in V \times \mathbb{R}^n \setminus 0 \times \Gamma_{N_o}$, where ξ is non-parallel to N and $N_o = d\phi(x_o)$.

(H.1) For $i \neq j$ $|\lambda_j(x,\xi,N) - \lambda_i(x,\xi,N)| \geq c_o > 0$

$\forall\ (x,\xi,N) \in V \times S^{n-1} \times \Gamma_{N_o}$.

(H.2) In $V \times S^{n-1} \times \Gamma_{N_o}$ the roots λ_j are either real or $|\operatorname{Im} \lambda_j| \geq c_o > 0$.

(H.3) The real roots are simple
The multiplicity of the complex roots is at most 3.

(H.4) p_m has C^∞ coefficients, p_{m-1} has Lipschitz coefficients and the others are bounded.
Then there exists a neighborhood W of the origin in which every $u \in C^m(V)$, such that $Pu = 0$ in V, $u = 0$ in $\{\phi(x) \leq \phi(x_o)\}$, vanishes.

For a proof of this result we refer to the original work of K. Watanabe [57] .

§4. A COUNTEREXAMPLE IN HIGER MULTIPLICITY

In this section we give a counterexample to the uniqueness for elliptic operators which proves that, when the characteristic roots are greater than three, conditions on the lower order terms are necessary in order to get uniqueness.

Theorem 4.1 :

Let a and b be positive integers such that $a \geq b$ and $2b-a-3 > 0$. We can find integers α, β , a complex number z and two $C^\infty(\mathbb{R}^2_{(x,t)})$ functions c,u vanishing for $t<0$ such that

$$(P+c)u = [\,(D_x - iD_t)^a + t^\alpha(1+zt^\beta)D_x^b - D_x^{b-1}\,]u + cu = 0$$

and $(0,0) \in \operatorname{supp} u$.

The admissible values of α,β,z will be given at the end of the proof. It will turn out that Theorem 4.1 applies to the following operators

$$P_1 = (D_x - iD_t)^4 + t^8(1-t^3)D_x^4 - D_x^3$$

$$P_2 = (D_x - iD_t)^6 + t^6(1-t^3)D_x^5 - D_x^4 .$$

Proof of Theorem 4.1 :

Since we use the same proof as in theorem 3.1 Chapter One, we shall omit some details and refer the reader to that proof.

Let us set $t - \delta = \frac{s}{\lambda}$ with $\lambda = \delta^{-\theta}$; then

$$\tilde{P} = (D_x - i\lambda D_s)^a + (\delta + \tfrac{s}{\lambda})^\alpha (1 + z(\delta + \tfrac{s}{\lambda})^\beta) D_x^b - D_x^{b-1} .$$

Setting

$$u(x,\delta,s) = e^{i\delta^{-\alpha}x + \frac{s}{\lambda}\delta^{-\alpha}} v_\delta(s)$$

and

$$I = e^{-\delta^{-\alpha}(ix + \frac{s}{\lambda})} \tilde{P} u$$

we get

$$(4.1) \quad I = [(-i\lambda D_s)^a + (\delta + \tfrac{s}{\lambda})^\alpha (1 + z(\delta + \tfrac{s}{\lambda})^\beta)\delta^{-\alpha b} - \delta^{-\alpha(b-1)}] v_\delta(s).$$

It is easy to see that

$$I = [(-i\lambda D_s)^a + \alpha s \delta^{\alpha - \alpha b - 1 + \theta} + z\delta^{\alpha + \beta - \alpha b} +$$

$$+ z(\alpha+\beta) \frac{s}{\lambda} \delta^{\alpha+\beta-\alpha b-1} + R] v_\delta(s)$$

where

$$R = \sum_{j=2}^{\alpha} \binom{\alpha}{j} s^j \delta^{\alpha-j-\alpha b+j\theta} + z \sum_{k=2}^{p} \binom{\alpha+\beta}{k} s^k \delta^{\alpha+\beta-k-\alpha b+k\theta} .$$

We take $\alpha - \alpha b - 1 + \theta = \alpha + \beta - \alpha b$, which means,

$$(4.2) \qquad \theta = 1 + \beta, \ \beta > 0 .$$

It follows that

$$I = [\,(-i\lambda D_s)^a + z\delta^{\beta+\alpha(1-b)} + f(\delta).s\ \delta^{\beta+\alpha(1-b)} + R\,]\,v_\delta(s)$$

where $f(\delta) = \alpha + \delta^{\theta-1} z(\alpha+\beta)$, so $f(0) = \alpha$.

Let us now look at R. By (4.2) we get since $j \geqslant 2$, $k \geqslant 2$,

$$\alpha - j - \alpha b + j\theta = \alpha(1-b) + j\beta > \alpha(1-b) + \beta$$

$$\alpha + \beta - k - \alpha b + k\theta = \alpha(1-b) + \beta(k+1) > \alpha(1-b) + \beta$$

It follows that we have $R = s^2\delta^{\alpha(1-b)+\beta}\psi(s,\delta)$ where ψ is a polynomial in s and δ. Now

$$I = [\,(-i\lambda D_s)^a + z\delta^{\beta+\alpha(1-b)} + \delta^{\beta+\alpha(1-b)} f(\delta)s + \delta^{\beta+\alpha(1-b)}\psi(s,\delta)s^2\,]\,v_\delta(s).$$

We take now $v_\delta(s) = e^{\nu\phi(s,\delta)} w(s,\delta)$ where ν is such that $(\lambda\nu)^a = \delta^{\beta+\alpha(1-b)}$ and $\nu \to \infty$ when $\delta \to 0$. This means, by (4.2), that

$$(4.3)\qquad \begin{cases} \nu = \delta^{\theta + \frac{1}{a}(\beta+\alpha(1-b))} \\[2ex] a(\beta+1) + \beta + \alpha(1-b) < 0 . \end{cases}$$

The function ϕ will be the solution of

$$(4.4)\qquad \begin{cases} (-\lambda\nu \dfrac{\partial\phi}{\partial s})^a + z\ \delta^{\beta+\alpha(1-b)} + \delta^{\beta+\alpha(1-b)} g(s,\delta)s = 0 \\[2ex] \phi_{|s=0} = 0 \end{cases}$$

where $g(s,\delta) = f(\delta) + \psi(s,\delta)s$, so $g(0,0) = \alpha$. By (4.3), equation (4.4) reduces to

$$(4.5)\qquad (-\frac{\partial\phi}{\partial s})^a + z + g(s,\delta).s = 0$$

so

$$(4.6)\qquad (\frac{\partial\phi}{\partial s})^a_{|s=0} = (-1)^{a+1} z .$$

Taking the derivative of (4.5) with respect to s and using (4.6) we get

(4.7) $$\frac{\partial^2\phi}{\partial s^2}\Big|_{s=0} = \frac{\alpha}{az}\left(\frac{\partial\phi}{\partial s}\right)\Big|_{s=0} .$$

(4.8) "If a is even we take $z = -1$".

We also take $\frac{\partial\phi}{\partial s}\Big|_{s=0} = 1$ so $\frac{\partial^2\phi}{\partial s^2}\Big|_{s=0} = -\frac{\alpha}{a}$.

It follows that

(4.9) $$\phi(s,\delta) = s - f(s,\delta)s^2$$

where $f \in B^\infty = \{\xi(s,\delta) \in C^\infty$ in s and $|D_s^\gamma \xi(s,\delta)| \leq c_\gamma, \delta \to 0\}$

and $f(0,0) = \frac{\alpha}{a} > 0$.

(4.10) "If a is odd we take $z = i$".

We also take for $\frac{\partial\phi}{\partial s}\Big|_{s=0}$ the value of $i^{1/a}$ having positive real part and negative imaginary part, which is possible since $a \geq 5$. It follows from (4.7) that

$$\mathrm{Re}\left(\frac{\partial^2\phi}{\partial s^2}\Big|_{s=0}\right) = \mathrm{Re}\left(-\frac{\alpha}{a}\, i\, \left(\frac{\partial\phi}{\partial s}\Big|_{s=0}\right)\right) = \frac{\alpha}{a}\, \mathrm{Im}\left(\frac{\partial\phi}{\partial s}\Big|_{s=0}\right) < 0 .$$

So we can write

(4.11) $$\begin{cases} \mathrm{Re}\ \phi(s,\delta) = d.s - h(s,\delta)s^2 \\ d > 0,\ h(0,0) > 0 . \end{cases}$$

Now by (4.1) and (4.4) it is easy to see that

$$I = \lambda^a \sum_{j=1}^{a} \nu^{a-j} L_j(s,\delta;\frac{\partial}{\partial s})\ w(s,\delta)$$

where L_j are operators of order $\leq j$ with coefficients in B^∞.

Moreover $L_1 = f_1(s,\delta)\frac{\partial}{\partial s}$ with $f_1(0,0) = (-1)^a\left(\frac{\partial\phi}{\partial s}\Big|_{s=0}\right)^{a-1} \neq 0$.

We can also write

(4.12) $$I = \lambda^a\ \nu^{a-1}\{L_1 w + \frac{1}{\nu}\ Qw\}$$

where Q is a differential operator of order $\leqslant a$ whose coefficients are in B^{∞}.
Let $w_j(s,\delta)$, $j \in \mathbb{N}$, be the solutions of

$$(4.13) \qquad \begin{cases} L_1 w_o = 1, \; w_{o|s=0} = 1 \\ L_1 w_j = -Qw_{j-1}, \; w_{j-1|s=0} = 0 \quad j \geqslant 1. \end{cases}$$

There exists a function $\psi(s,\delta,r)$ in B^{∞} with respect to (s,δ) such that

$$|D_s^{\ell}(\psi - \sum_{j<N} r^j w_j(s,\delta))| \leqslant C\, r^N$$

for (s,δ,r) near $(0,0,0)$. We shall take $w(s,\delta) = \psi(s,\delta,\frac{1}{\nu})$ where ν is given by (4.3).
Now setting

$$(4.14) \qquad v_{\delta}(x,s) = e^{i\delta^{-\alpha}x + \delta^{-\alpha}\frac{s}{\lambda}}\, e^{\nu\phi(s,\delta)} w(s,\delta)$$

we have

Lemma 4.2 :
Let us set $r_{\delta} = \dfrac{\tilde{P}v_{\delta}}{v_{\delta}}$. Then for all integers p,q there exists $\mathcal{H}$ such that for every $N \in \mathbb{N}$ there exists $C = C(p,q,N)$ such that when $\delta \to 0$,

$$|D_x^p\, D_s^q\, r_{\delta}| \leqslant C\, \delta^{\mathcal{H}N}$$

Proof :
See pronosition 3.4, Chapter One.

In the following, k will be an integer going to $+\infty$. We set

$$(4.15) \qquad \begin{cases} \delta = b_k = k^{-\rho} \\ \rho = \frac{1}{\beta}. \end{cases}$$

With this choice, $|s| = |\frac{t-b_k}{b_k^{\theta}}| \leqslant \rho k^{-\rho-1} k^{\rho\theta} = \rho$.

We are going now to modify slightly v_k given by (4.14) into functions $\tilde{v}_k$ such that $\tilde{r}_k = \dfrac{\tilde{P}\tilde{v}_k}{\tilde{v}_k}$ satisfies Lemma 4.2 and moreover, setting $m_k = \frac{1}{2}(b_k + b_{k+1})$, $\tilde{r}_k$ is flat at the points m_{k-1} and m_k.

Let us set

$$(4.16)\quad \tilde{v}_k(x,t) = e^{ib_k^\alpha x + (t-b_k)b_k^{-\alpha}}\, e^{\nu_k \phi(\frac{t-b_k}{b_k^\theta}, b_k)} \{w_k(\frac{t-b_k}{b_k^\theta}) + f_k(\frac{t-b_k}{b_k^\theta})\}$$

where $f_k(s) = f(s,b_k)$ are to be chosen.

It is easy to see that

$$\tilde{r}_k = \frac{\tilde{P}\tilde{v}_k}{\tilde{v}_k} = \frac{r_k w_k}{w_k+f_k} + k^\mu \frac{\nu_k^{-1} Qf_k + L_1 f_k}{w_k + f_k}$$

where $k^\mu = \lambda_k^a \nu_k^{a-1}$ and Q(resp. L_1) is an operator of order $\leqslant a$ (resp. $\leqslant 1$) with respect to $\partial/_{\partial s}$ whose coefficients are in B^∞. Let us take f_k such that

$$(\frac{\partial}{\partial t})^\ell f_k|_{t=m_k} = 0,\ 0 \leqslant \ell \leqslant a-1.$$

It is easy to see (see step 6, theorem 3.1 Chap.1) that asking $\tilde{r}_k$ to be flat on $t = m_k$ is equivalent to imposing prescribed values of all the derivatives of f_k at the point $t = m_k$ (the same being true for $t = m_{k-1}$) which are rapidly decreasing with respect to k . By the well known theorem of Borel, such an f_k exists satisfying

$|D_s^\alpha f_k| \leqslant C_{\alpha,N}\, k^{-N}$ for all N .

Let us now set

$$(4.17)\qquad \tilde{F}_k(x,t) = \mathrm{Log}\left|\frac{\tilde{v}_k}{\tilde{v}_{k+1}}\right| .$$

As in Th. 3.1 Chap. 1, we want this function to be increasing.
By (4.9), (4.11) and (4.16) it is easy to see that :

$$\frac{\partial \tilde{F}_k}{\partial t} = \underbrace{k^{\alpha\rho} - (k+1)^{\alpha\rho}}_{①} + \beta_o\underbrace{\left(\frac{\nu_k}{b_k^\theta} - \frac{\nu_{k+1}}{b_{k+1}^\theta}\right)}_{②} - \frac{\nu_k}{b_k^\theta}\,.\,\underbrace{\beta_1(t_k, b_k)\,\frac{t-b_k}{b_k^\theta}}_{③} +$$

$$+ \underbrace{\frac{\nu_{k+1}}{b_{k+1}^\theta}\,\beta_1(t_{k+1},}\, b_{k+1})\,\frac{t-b_{k+1}}{b_{k+1}^\theta} + O(1)$$

where $\beta_1(s,\delta)$ is in B^∞ with $\beta_1(0,0) > 0$, $\beta_o > 0$ and $t_k = \frac{t-b_k}{b_k^\theta}$. Now we have

$$(4.18) \qquad ① \sim -\alpha\rho\; k^{\alpha\rho-1}$$

$$(4.19) \qquad ② \sim -\, c_o\, k^{-\frac{\rho}{a}(\beta+\alpha(1-b))-1}, \quad c_o > 0\,.$$

On the other hand, the main contribution in the third term ③ is obtained by taking $\beta_1(0,0)$. We get

$$③ = \beta_1(0,0)\;\underbrace{\frac{\nu_{k+1}}{b_{k+1}^{2\theta}}(b_k - b_{k+1})}_{I/} + \underbrace{(t-b_k)\left\{\frac{\nu_{k+1}}{b_{k+1}^{2\theta}} - \frac{\nu_k}{b_k^{2\theta}}\right\}}_{II/}\,.$$

It is easy to see that

$$I/ \geqslant c_1\; k^{\rho(\beta+1) - \frac{\rho}{a}(\beta+\alpha(1-b)) - \rho - 1}, \quad c_1 > 0$$

$$II/ \sim c_2\; k^{-\rho-1}\; k^{\rho(\beta+1)}\; k^{-\frac{\rho}{a}(\beta+\alpha(1-b))-1}\,.$$

Since $\rho\beta = 1$, we get for $k \to \infty$,

$$(4.20) \qquad ③ \geqslant c_3\; k^{-\frac{\rho}{a}(\beta+\alpha(1-b))}, \quad c_3 > 0\,.$$

In order to get an increasing $\tilde{F}_k$ it is necessary that ③ dominates ① and ② given by (4.18) and (4.19). We shall take

$$-\frac{\rho}{a}(\beta+\alpha(1-b)) > \alpha\rho-1$$

which is equivalent, since $\rho\beta = 1$, to

(4.21) $$\alpha(a+1-b) < \beta(a-1).$$

With this choice, we have when $k \to \infty$

(4.22) $$\frac{d\tilde{F}_k}{dt} \geqslant c_4 \, k^{-\frac{\rho}{a}(\beta+\alpha(1-b))}, \quad c_4 > 0.$$

Let us note that $\beta + \alpha(1-b) < 0$.

Let $m_k = \frac{1}{2}(b_k + b_{k+1})$. We are going to prove that we can find a sequence of positive numbers c_k such that if we set

(4.23) $$u_k(x,t) = e^{-c_k}\,\tilde{v}_k(x,t)$$

where $\tilde{v}_k$ is given by (4.16), then

(4.24) $$|u_k| = |u_{k+1}| \quad \text{for} \quad t=m_k$$

(so $|u_k| = |u_{k-1}|$ at $t=m_{k-1}$).

Let us remark that if we set $G_k(x,t) = \mathrm{Log}\left|\frac{u_k}{u_{k+1}}\right|$ then $\frac{dG_k}{dt} = \frac{dF_k}{dt}$, so G_k also satisfies (4.22).

On the other hand, $\frac{Pu_k}{u_k} = \frac{P\tilde{v}_k}{\tilde{v}_k}$ so $\frac{Pu_k}{u_k}$ satisfies Lemma 4.2 and is flat at $t=m_k$ and $t=m_{k-1}$.

Taking the Logarithm of (4.24) and using (4.16) we see that we must have

$$\underbrace{c_{k+1} - c_k}_{①} + \underbrace{k^{\alpha\rho}(m_k - b_k) - (k+1)^{\alpha\rho}(m_k - b_{k+1})}_{②} +$$

$$(4.25) \qquad + \nu_k \operatorname{Re}\phi\left(\frac{m_k - b_k}{b_k^\theta}, b_k\right) - \underbrace{\nu_{k+1} \operatorname{Re}\phi\left(\frac{m_k - b_{k+1}}{b_{k+1}^\theta}, b_{k+1}\right)}_{③} +$$

$$+ \operatorname{Log} \left| \underbrace{\frac{w_k + f_k}{w_{k+1} + f_{k+1}}}_{④} \right|_{t=m_k} = 0 .$$

We have

$$(4.26) \qquad ② = -\frac{1}{2}(b_k - b_{k+1})(k^{\alpha\rho} + (k+1)^{\alpha\rho}) \sim -\rho\, k^{\alpha\rho - \rho - 1} .$$

Now, since $\operatorname{Re}\phi = \beta_o s - \beta_1(s,\delta)s^2$ with $\beta_o > 0$, $\beta_1(0,0) > 0$, we get

$$③ = \nu_k\left\{\beta_o \frac{m_k - b_k}{b_k^\theta} - \beta_1\left(\frac{m_k - b_k}{b_k^\theta}, b_k\right)\left(\frac{m_k - b_k}{b_k^\theta}\right)^2\right\} - \nu_{k+1}\left\{\frac{m_k - b_{k+1}}{b_{k+1}^\theta}.\beta_o - \right.$$

$$\left. - \beta_1(\ldots)\left(\frac{m_k - b_{k+1}}{b_{k+1}^\theta}\right)^2\right\} .$$

Let us set $I = \beta_o \nu \dfrac{m_k - b_k}{b_k^\theta} - \beta_o \nu_{k+1} \dfrac{m_k - b_{k+1}}{b_{k+1}^\theta}$ and $II = ③ - I$.

We have

$$I = \beta_o \frac{\nu_k}{b_k^\theta}(m_k - b_k - m_k + b_{k+1}) + \beta_o(m_k - b_{k+1}).\left\{\frac{\nu_k}{b_k^\theta} - \frac{\nu_{k+1}}{b_{k+1}^\theta}\right\}.$$

The first term of this equality is equivalent to

$k^{-\rho-1}\, k^{-\frac{\rho}{a}(\beta + \alpha(1-b))}$ by (4.3). The second one is equivalent to

$k^{-\rho-1}\, k^{-\frac{\rho}{a}(\beta + \alpha(1-b)) - 1}$, so

$$(4.27)\qquad I \sim c\, k^{-\rho-1-\frac{\rho}{a}(\beta+\alpha(1-b))} .$$

Let us consider II. It is easy to see that the main term comes from $\beta_1(0,0)$. Since $(m_k-b_k)^2 = (m_k-b_{k+1})^2$ we can write

$$II \sim (m_k-b_k)^2\ \beta_1(0,0)\{\frac{\nu_k}{b_k^{2\theta}} - \frac{\nu_{k+1}}{b_{k+1}^{2\theta}}\} .$$

It follows that

$$(4.28)\qquad II \sim c' k^{-2\rho-2}\, k^{-\frac{\rho}{a}(\beta+\alpha(1-b))-1}\, k^{-\rho(\beta+1)} .$$

From (4.27) and (4.28) we see that I is the main term. Indeed,

$$-\frac{\rho}{a}(\beta+\alpha(1-b))-\rho-1 > -2\rho-2-\frac{\rho}{a}(\beta+\alpha(1-b))-1-\rho\beta-\rho$$

since $\rho\beta=1$. We get

$$(4.29)\qquad ③ \sim c\, k^{-\frac{\rho}{a}(\beta+\alpha(1-b))-\rho-1} .$$

It follows from (4.26) and (4.29) that ② dominates ③. Indeed, the inequality

$$\alpha\rho-\rho-1 > -\frac{\rho}{a}(\beta+\alpha(1-b))-\rho-1$$

is equivalent to $a\alpha+\beta > b\alpha-\alpha$, which is true since $a \geqslant b$ and $\beta>0$. Finally, it is easy to see that ④ $= O(1)$. From (4.25) we get

$$c_{k+1} - c_k - \rho\, k^{\rho(\alpha-1)-1}(1+o(1)) = 0 .$$

The c_k are then defined by induction. Moreover we get

$$(4.30)\qquad c_k \sim c_1 k^{\rho(\alpha-1)},\ \alpha>1 .$$

Summing up, we have proved that, setting

$$(4.31)\qquad u_k(x,t) = e^{-c_k}\, e^{ib_k^{-\alpha}x+(t-b_k)b_k^{-\alpha}}\, e^{\nu_k\phi(\frac{t-b_k}{b_k^{\theta}},\, b_k)}\{w_k+f_k\} ,$$

we have

$$(4.32)\quad \begin{cases} |u_k| = |u_{k+1}| \quad \text{at} \quad t=m_k \\ r_k = \dfrac{\tilde{P}u_k}{u_k} \text{ is flat at } t=m_k,\ t=m_{k-1} \text{ and is} \\ \text{rapidly decreasing in } k \\ \dfrac{d}{dt}\left(\text{Log}\left|\dfrac{u_k}{u_{k+1}}\right|\right) \geqslant c\, k^{-\frac{\rho}{a}(\beta+\alpha(1-b))},\quad c>0,\ k\to\infty . \end{cases}$$

Let us recall that we must have by (4.3), (4.15), (4.21)

$$(4.33)\quad \begin{cases} a(\beta+1) + \beta + \alpha(1-b) < 0 \\ \alpha(a+1-b) < \beta(a-1),\ \rho = \dfrac{1}{\beta} \end{cases}$$

and it follows from (4.3) that

$$(4.34)\quad -\frac{\rho}{a}(\beta+\alpha(1-b)) > \frac{\rho}{a}\, a(\beta+1) = \rho+1 .$$

Now we see easily, using (4.2), (4.3) and (4.30), that $\frac{\nu_k}{c_k} = O(\frac{1}{k})$. On the other hand, $|t-b_k|b_k^{-\alpha} \leqslant c\, k^{\rho\alpha-\alpha-1} = c_k O(\frac{1}{k})$. So all the derivatives of u_k, defined by (4.31), tend to zero when k goes to $+\infty$. Let χ be a C^∞ function on the line such that $\chi(s) = 1$ for $|s| \leqslant 3/4$, supp $\chi \subset [-1,1]$, $0 \leqslant \chi \leqslant 1$, and define

$$u(x,t) = \begin{cases} \sum_{k\geqslant k_o} \chi_k(t)\, u_k(x,t),\ 0<t\leqslant b_{k_o} \\ 0 \qquad\qquad\qquad\qquad ,\ t\leqslant 0 \end{cases}$$

with $\chi_k(t) = \chi(\frac{t-b_k}{\ell_k})$, $\ell_k = b_k - b_{k+1}$.

Then using (4.32), (4.34) and exactly the same method as in Step 7, theorem 3.1 Chapter Cne, we can prove that $a = -\frac{Pu}{u}$ is C^∞ near the origin and flat at $t=0$.

Let us check now the compatibility of (4.33). The first inequality gives $\alpha > \beta \frac{a+1}{b-1} + \frac{a}{b-1}$; the second one implies $\alpha < \beta \frac{a-1}{a+1-b}$. So we can find α only if $\beta \frac{a+1}{b-1} + \frac{a}{b-1} <$ $< \beta \frac{a-1}{a+1-b}$, which is equivalent to

$$\beta \left\{ \frac{a-1}{a+1-b} - \frac{a+1}{b-1} \right\} > \frac{a}{b-1} .$$

If

$$\frac{a-1}{a+1-b} - \frac{a+1}{b-1} > 0 \tag{4.35}$$

we can take

$$\beta > \frac{a}{b-1} \left\{ \frac{a-1}{a+1-b} - \frac{a+1}{b-1} \right\}^{-1} = \frac{a+1-b}{2b-a-3} \tag{4.36}$$

$$\beta \frac{a+1}{b-1} + \frac{a}{b-1} < \alpha < \beta \frac{a-1}{a+1-b} . \tag{4.37}$$

Now (4.35) is clearly equivalent to $2b-a-3 > 0$, which was assumed in Theorem 4.1.

If $b=a-1$ then $a > 5$. We can take $a=6$, $b=5$, $\beta=3$, $\alpha=6$.
If $b=a$ then $a > 3$ and we can take $a=4$, $b=4$, $\beta=3$, $\alpha=8$.

§5. UNIQUENESS THEOREMS IN CASE OF HIGHER MULTIPLICITY

We state in this section (without proof) some results giving sufficient conditions for uniqueness, in case of operators having roots of arbitrary high multiplicity. The first one concerns elliptic operators, but it can be easily extended to operators having also real roots of multiplicity one.

Let $P(x,D)$ be an elliptic differential operator of order m in a neighborhood V of the origin in $\mathbb{R}^n$ with C^∞

coefficients. We shall denote by $p(x,\xi)$ (resp. $p_m(x,\xi)$) its symbol (resp. principal symbol). The initial hypersurface S will be given by a C^2 real function f such that $\operatorname{grad} f(0) = N_o \neq 0$.

We shall denote by r the maximum multiplicity of the roots τ of the equation $p_m(0;\xi+\tau N_o) = 0$ where $\xi \in \mathbb{R}^n$ is non-parallel to N_o. We set $q = \left[\frac{r+1}{2}\right]$. Now $S(p_m,N_o)$ will denote the set of polynomials $a(x,\xi)$ in ξ such that

$$a(x,\xi) = \sum_{|\alpha|\leqslant m-q} b_\alpha(x)\xi^\alpha + \sum_{|\lambda|\geqslant 1,\, 2|\beta|+|\mu|\leqslant|\lambda|} c_{\beta,\mu,\lambda}(x)\xi^\beta\, p^{(\lambda)}_{m(\mu)}(x,\xi)$$

where $p^{(\lambda)}_{m(\mu)} = D^\mu_x\left(\frac{\partial}{\partial\xi}\right)^\lambda p_m$ and b_α, $c_{\beta,\mu,\lambda} \in L^\infty(V)$.

We can state the

Theorem 5.1 :

Let $P(x,D)$ be an elliptic differential operator with C^∞ coefficients in a neighborhood V of the origin in $\mathbb{R}^n$. Let us assume that

(H.1) *For every $\xi_o \in \mathbb{R}^n$ non-parallel to N_o, there exists m C^∞ functions $\lambda_j(x,\xi,N)$ in a conic neighborhood of $(0,\xi_o,N_o)$ such that*

$$p_m(x,\xi+\tau N) = p_m(x,N)\prod_{j=1}^m(\tau-\lambda_j(x,\xi,N))$$

(H.2) $p - p_m \in S(p_m,N_o)$.

Then there exists a neighborhood W *of the origin such that if* u *is a* C^m *function in* V *satisfying* $P(x,D)u=0$ *in* V *and* $u=0$ *in* $\{x \in V : f(x) \leq f(0)\}$, *then* u *vanishes in* W.

Remarks 5.2:

(1) This theorem contains, in the elliptic case, theorem 1.1 of this chapter, since for $r \leq 2$ (H.2) is always satisfied.

(2) Let us consider the following symbol : $p_m = (\xi - i\tau)^4 + t^8(1-t^3)\xi^4$. It is easily seen that (H.1) is satisfied. Then according to the definition of $S(p_m, N_o)$ we see that $b(x,t)t^7 . \xi^3 + \sum_{|\alpha| \leq 2} a_\alpha(x,t)\xi^\alpha$ belongs to $S(p_m, N_o)$ for all a_α and b in $L^\infty(V)$. So theorem 5.1 implies uniqueness for

$$P = (D_x - iD_t)^4 + t^8(1-t^3)D_x^4 + b(x,t)t^7 D_x^3 + \\ + \sum_{|\alpha| \leq 2} a_\alpha(x,t) D_x^\alpha .$$

This example must be compared with that given after theorem 4.1. The following result concerns, in case of constant multiplicity, operators having real and complex roots of arbitrary high multiplicities. We begin by the elliptic case.

Theorem 5.2:

Let $P(x,D)$ *be an elliptic differential operator in a neighborhood of a point* $x_o \in \mathbb{R}^n$ *and* $S = \{\phi(x) = \phi(x_o)\}$ *a* C^2 *non-characteristic hypersurface. Denoting* $N_o = \text{grad}\, \phi(x_o)$, *we suppose that*

$$\text{(H.1)}\quad \begin{cases} \exists V_{x_0},\ \exists \Gamma_{N_0} : \ \forall N \in \Gamma_{N_0},\ \forall \xi \ \underline{\text{non-parallel to}}\ N \\ \underline{\text{there exist}}\quad \lambda_j \in C^\infty(V_{x_0} \times \mathbb{R}^n \setminus 0 \times \Gamma_{N_0})\ j=1,\dots,k \\ \underline{\text{such that}} \\ p_m(x,\xi+\tau N) = a(x) \prod_{j=1}^{k} (\tau-\lambda_j(x,\xi,N))^{m_j}. \end{cases}$$

(H.2) $\exists c_0 > 0 : |\lambda_j(x,\xi,N)-\lambda_\ell(x,\xi,N| \geq c_0$

$$\forall (x,\xi,N) \in V_{x_0} \times S^{n-1} \times \Gamma_{N_0},\ j \neq \ell.$$

(H.3) $\forall j=1,\dots,k,\ m_j \geq 2$ and $p^s_{m-1}(x,\xi+\tau N)|_{\tau=\lambda_j(x,\xi,N)} \neq 0$

for $(x,\xi,N) \in V \times \mathbb{R}^n \setminus 0 \times \Gamma_{N_0}$.

(H.4) The coefficients of p_m and p_{m-1} are C^∞, the others are bounded.

Then there exists a neighborhood W of x_0 in which every solution $u \in C^m$ of

$$\begin{cases} Pu = 0 \ \text{in}\ V \\ u = 0 \ \text{in}\ \{\phi(x) < \phi(x_0)\} \end{cases}$$

vanishes.

In the case of real and complex roots we can state

Theorem 5.3 :

Let us suppose that p_m has real coefficients, that, with the notations of theorem 5.2, the conditions (H.1), (H.2) and (H.4) are satisfied and furthermore that

(H.3) $\forall j=1,\dots,k,\ m_j \geq 2$ and $\operatorname{Im} p^s_{m-1}|_{\tau=\lambda_j} \neq 0$ for every

$(x,\xi,N) \in V_{x_o} \times \mathbb{R}^n \setminus 0 \times \Gamma_{N_o}$.

Then the same conclusion as in theorem 5.2 holds.

(In these two results p^s_{m-1} denotes the subprincipal symbol of P).

Let us remark that (H.3) cannot be replaced by $p^s_{m-1}|_{\tau=\lambda_j} \neq 0$. Indeed, let us consider in $\mathbb{R}^2$ the operator $D_t^2 + D_x$; it is easy to see that it is not stronger than D_x, so by theorem 5.1 Chapter 3, we can find $a=\alpha+i\beta$ flat at $t=0$ such that

$$D_t^2 + D_x + (\alpha+i\beta)D_x = D_t^2 + ((\alpha+1) + i\beta)D_x$$

does not possess the uniqueness property. Now $p^s_{m-1}(0,0;\xi,\tau)|_{\tau=0}=\xi\neq 0$. Finally we give a result which shows that the subprincipal symbol can be replaced by symbols of lower order.

Theorem 5.4 :

Let, in $\mathbb{R}^{n+1} = \{(x,t) : x\in\mathbb{R}^n, t\in\mathbb{R}\}$, P be defined by

$$P = D_t^m + \sum_{\ell=1}^{m} a_\ell(x,t)D_t^{m-\ell} + \sum_{|\alpha|+\beta\leq m-j} a_{\alpha\beta}(x,t)D_x^\alpha D_t^\beta$$

where j is an integer such that $1 \leq j < m$.

Let us suppose that

$$\text{(H.1)}\quad \begin{cases} a_\ell \in L^\infty,\ 1\leq \ell\leq m\ ;\ a_{\alpha\beta}\in L^\infty,\ |\alpha|+\beta < m-j \\ a_{\alpha\beta} \in C^\infty \quad |\alpha|+\beta=m-j\ . \end{cases}$$

(H.2) $\begin{cases} \underline{\text{If } m \text{ is even}}: \forall \xi \in S^{n-1}, \sum_{|\alpha|=m-j} a_{\alpha,0}(0,0)\xi^\alpha \notin \mathbb{R}_- = \\ = \{\lambda \in \mathbb{R} ; \lambda \leq 0\} . \\ \\ \underline{\text{If } m \text{ is odd}}: \forall \xi \in S^{n-1}, \sum_{|\alpha|=m-j} a_{\alpha,0}(0,0)\xi^\alpha \notin \mathbb{R} . \end{cases}$

Then there exists a neighborhood of the origin in which every $u \in C^m$ satisfying

$$\begin{cases} Pu = 0 \\ u|_{t<0} = 0 \end{cases}$$

vanishes.

Let us note that theorem 5.3 is false if there exists $\xi_o \in S^{n-1}$ such that $\sum_{|\alpha|=m-j} a_{\alpha,0} \xi_o^\alpha \in \mathbb{R}_-^\star$, when m is even (or $\mathbb{R}^*$ when m is odd).

COMMENTS

(1) Results like theorem 1.1 have been initiated by A.P. Calderon in [8], and his first result has been improved by several people. The case of variable multiplicity given here and its proof follow closely K. WATANABE - C. ZUILY [59] (See also [62] for related results). However the Carleman estimates for elementary operators are taken from L. NIRENBERG [35], and we refer to that paper for an improvement concerning the double roots. The smoothness of the coefficients assumed in theorem 1.1 is rather strong and we refer to [18] for uniqueness results under minimal smoothness assumptions.

(2) Example (2.1) given at the beginning of §2 is due to

A. PLIŠ [42] . See also theorem 4.4 Chapter 3 for a general result containing this example.

(3) Theorem 2.1 is due to C. ZUILY [65] . See also K. WATANABE [58] . A more general result concerning higher order operators and including real roots can be found in H. OUNAIES [37].

(4) Example (3.1) is due to P. COHEN [12]. Theorem 3.1 is taken from K. WATANABE [57].

(5) Theorem 4.1 is a result of A. PLIŠ [42]. However the proof given here follows L. HÖRMANDER [20].

(6) Theorem 5.1. is due to K. WATANABE - C. ZUILY [59]; it unifies and extends several previous results of L. HÖRMANDER [18], P.M. GOORJAN [16], S. MIZOHATA [32] etc...
Theorems 5.2 and 5.3 are results of C. ZUILY [64]. They extend previous results of W. MATSUMOTO [30] (see also M. ZEMAN [63], but it seems that there is a gap in his proof). Finally, theorem 5.3 is taken from M. DAMLAKHI - C. ZUILY [13] .

Chapter Three

UNIQUENESS AND PSEUDO-CONVEXITY

In the results described before, no condition (except to be non-characteristic) was imposed on the initial hypersurface and in particular, uniqueness did not depend on the side containing the support of the solution ; however, precise hypotheses like smoothness, multiplicity of the characteristic roots, were made. In the general case (where no smoothness occurs) we shall see that uniqueness depends on geometrical conditions between the operator and the hypersurface, called "pseudo-convexity conditions".

§0. PRELIMINARIES

0.1. We shall at first recall some facts concerning the Weyl calculus of (pseudo) differential operators and we refer the reader to [21] for further developments of the theory.

Let $P(x,D)$ be a differential operator with usual symbol $\tilde{p}(x,\xi)$; then its Weyl symbol $p(x,\xi)$ is given by

$$p(x,\xi) = \sum_{k \geqslant 0} \left(-\frac{1}{2}\right)^k \frac{1}{k!} (D_\xi \partial_x)^k \tilde{p}(x,\xi)$$

where

$$D_\xi \partial_x = \frac{1}{i} \sum_{j=1}^{n} \frac{\partial^2}{\partial x_j \partial \xi_j} .$$

Then, for $u \in \mathcal{S}$,

$$(1.1) \qquad Pu(x) = (2\pi)^{-n} \iint e^{i \langle x-y,\xi \rangle} p\left(\frac{x+y}{2},\xi\right) u(y) dy \, d\xi .$$

Moreover ,

(1.2) The Weyl symbol of P^*, the adjoint of P, is $\overline{p}$.

The composition, denoted by #, is given by

(1.3) $$a \# b = \sum_{\alpha,\beta} \frac{1}{\alpha!} \cdot \frac{1}{\beta!} \left(\frac{1}{2}\right)^{|\alpha|} \left(-\frac{1}{2}\right)^{|\beta|} D_\xi^\alpha \partial_x^\beta a \, D_\xi^\beta \partial_x^\alpha b$$

where $D_{\xi_j} = \frac{1}{i} \frac{\partial}{\partial \xi_j}$ and $\partial_{x_j} = \frac{\partial}{\partial x_j}$.

Now let γ be a positive real number and ψ be a real C^∞ function in $\mathbb{R}^n$ such that

(1.4) $$D^\alpha \psi \equiv 0 \, , \, |\alpha| \geqslant 3 \, .$$

Let us set

(1.5) $$P_\gamma = e^{\gamma\psi} P e^{-\gamma\psi} \, .$$

We claim that

(1.6) The Weyl symbol of P_γ is $p(x, \xi + i\gamma\psi'(x))$, where $\psi'(x)$ stands for grad $\psi(x)$.

Indeed, we can write

$$P_\gamma u(x) = (2\pi)^{-n} \iint e^{i\langle x-y,\, \xi\rangle} p\left(\frac{x+y}{2}, \xi\right) e^{\gamma(\psi(x)-\psi(y))} u(y) dy d\xi .$$

By (1.4) we have

$$\psi(x) - \psi(y) = \psi'\left(\frac{x+y}{2}\right) . (x-y) \, ,$$

so

$$P_\gamma u(x) = (2\pi)^{-n} \iint e^{i\langle x-y,\, \xi - i\gamma\psi'(\frac{x+y}{2})\rangle} p\left(\frac{x+y}{2}, \xi\right) u(y) dy d\xi .$$

Since p is a polynomial in ξ,

$$P_\gamma u(x) = (2\pi)^{-n} \iint e^{i\langle x-y, \eta\rangle} p\left(\frac{x+y}{2}, \eta + i\gamma\psi'\left(\frac{x+y}{2}\right)\right) u(y) dy d\eta . \blacksquare$$

In the following we shall use a weight function ψ of the form

(1.7) $\psi(x) = \varphi(x_o) + \varphi'(x_o).(x-x_o) + \frac{1}{2}\varphi''(x_o).(x-x_o)^2 +$

$$+ \frac{C}{2}(\varphi'(x_o)(x-x_o))^2 - \frac{1}{2C^2}\,||x-x_o||^2$$

where φ is a C^2 function near a point $x_o \in \mathbb{R}^n$ and C a positive constant.

With this weight function, using (1.2), (1.3), (1.6) and (1.7), we see easily that we have the following formulas :

(1.8) $\sigma(P_\gamma^* P_\gamma) = |p_m(x,\zeta)|^2 + C\gamma\,|\frac{\partial p_m}{\partial \xi}(x,\zeta).\,\frac{\partial \varphi}{\partial x}(x_o)|^2 +$

$$+ \gamma\,\frac{\partial^2 \varphi}{\partial x^2}(x_o).\,\overline{\frac{\partial p_m}{\partial \xi}(x,\zeta)}\cdot\frac{\partial p_m}{\partial \xi}(x,\zeta) + \mathrm{Im}\,\overline{\frac{\partial p_m}{\partial \xi}(x,\zeta)}.\,\frac{\partial p_m}{\partial x}(x,\zeta) -$$

$$- \frac{\gamma}{C^2}\,\overline{\frac{\partial p_m}{\partial \xi}(x,\zeta)}.\frac{\partial p_m}{\partial \xi}(x,\zeta) + 2\,\mathrm{Re}\,\overline{p_m(x,\zeta)}.p_{m-1}(x,\zeta) +$$

$$+ \sum_{j \geqslant 2}\,\sum_{\ell=0}^{j} a_{2m-\ell-j}(x,\zeta)\,\gamma^\ell\,\Pi_\ell(C)\ .$$

Let us explain the notations used above : σ denotes the Weyl symbol of the operator under study, p_m is the principal symbol of P, $\zeta = \xi + i\gamma\psi'(x)$, $a_{2m-\ell-j}(x,\zeta)$ are homogeneous in ζ of order $2m-\ell-j$ and $\Pi_\ell(C)$ satisfies $\lim_{C\to\infty} C^{-\ell}\,\Pi_\ell(C) = \alpha_\ell \in \mathbb{C}$. Moreover $\frac{\partial p}{\partial \xi} = (\frac{\partial p}{\partial \xi_j})_{j=1,\dots,n}$;

if a,b are in $\mathbb{C}^n$, $a.b = \sum_{j=1}^{n} a_j b_j$ and $\frac{\partial^2\varphi}{\partial x^2}.a.b = \sum_{j,k}\frac{\partial^2\varphi}{\partial x_j\,\partial x_k}\,a_j\,b_k$.

By the same way we get

(1.9) $\sigma(P_\gamma P_\gamma^*) = |p_m(x,\zeta)|^2 - C\,\gamma\,|\frac{\partial p_m}{\partial \xi}(x,\zeta).\,\frac{\partial \varphi}{\partial x}\,(x_o)|^2 -$

$$-\gamma \frac{\partial^2\varphi}{\partial x^2}(x_o).\overline{\frac{\partial p_m}{\partial\xi}(x,\zeta)}.\frac{\partial p_m}{\partial\xi}(x,\zeta) - \operatorname{Im}\overline{\frac{\partial p_m}{\partial\xi}(x,\zeta)}.\frac{\partial p_m}{\partial x}(x,\zeta) +$$

$$+\frac{\gamma}{C^2}\overline{\frac{\partial p_m}{\partial\xi}(x,\zeta)}.\frac{\partial p_m}{\partial\xi}(x,\zeta) + 2\operatorname{Re}\overline{p_m(x,\zeta)}\ .\ p_{m-1}(x,\zeta) +$$

$$+\sum_{j\geqslant 2}\sum_{\ell=0}^{j}(-1)^\ell\, a_{2m-j-\ell}(x,\zeta)\ .\ \gamma^\ell\, \Pi_\ell(C)\ .$$

0.2 . Let us now introduce a class of pseudo-differential operators depending of the positive parameter γ.

Let Ω be an open set in $\mathbb{R}^n$ and $m\in\mathbb{R}$. S^m_γ will denote the set of functions $a(x,\xi,\gamma)$ which are C^∞ with respect to (x,ξ) in $\Omega\times\mathbb{R}^n$ and such that for every compact $K\subset\Omega$ and all $\alpha,\beta\in\mathbb{N}^n$, $\xi\in\mathbb{R}^n$, $\gamma\geqslant\gamma_o$,

$$\sup_K |D_x^\beta\, D_\xi^\alpha\, a(x,\xi,\gamma)|\leqslant C_{\alpha\beta K}(\gamma+|\xi|)^{m-|\alpha|}.$$

We shall denote by $OP(S^m_\gamma)$ the set of all pseudo-differential operators with Weyl symbol in S^m_γ .

A complete symbolic calculus, analogous to the usual one, can be developed for these classes and we shall give here the main points.

1°/ Let $A\in OP(S_\gamma^{m_1})$, $B\in OP(S_\gamma^{m_2})$ be two p.d.o. with Weyl symbol a and b. If A or B is proper then $A\ \#\ B$ is a p.d.o in $OP(S_\gamma^{m_1+m_2})$ and :

For every $N\in\mathbb{N}$ we can find $r\in S_\gamma^{m_1+m_2-N}$ such that

$$(1.10)\quad \sigma^W(A\ \#\ B) = \sum_{|\alpha|+|\beta|<N}\frac{1}{\alpha!}\cdot\frac{1}{\beta!}\left(\frac{1}{2}\right)^{|\alpha|}\left(-\frac{1}{2}\right)^{|\beta|} D_\xi^\alpha\,\partial_x^\beta a.\ D_\xi^\beta\,\partial_x^\alpha b + r\ .$$

In particular, if $N=2$, we have

(1.11) $$\sigma^W(A \# B) = a.b + \frac{1}{2i}\{a,b\} + r \ , \quad r \in S_\gamma^{m_1+m_2-2}$$

where $\{a,b\}$ is the Poisson bracket of a and b.

2°/ Let $A \in OP(S_\gamma^{2m})$, with Weyl symbol $a(x,\xi,\gamma)$, be such that :

(1.12) $$\exists\ \delta_o > 0,\ \gamma_o > 0 : \forall\ \xi \in \mathbb{R}^n,\ \forall\ \gamma \geqslant \gamma_o,\ a(x,\xi,\gamma) \geqslant \delta_o(|\xi| + \gamma)^{2m}.$$

Then

(i) $a^{1/2} \in S_\gamma^m$

(ii) $\exists\, B \in OP(S_\gamma^m),\ R \in OP(S_\gamma^{2m-2}) : A = B^* \# B + R$.

The first fact is obvious ; for the second one, let B be the p.d.o with Weyl symbol $a^{1/2}$; then B is self-adjoint and by (1.11) we have $\sigma(B \# B) = a + r$ where $r \in S_\gamma^{2m-2}$. It follows that we can find $M > 0$ such that for every $\gamma \geqslant \gamma_o$ and $u \in C_o^\infty$,

(1.13) $$(Au,u)_{L^2} \geqslant -M \cdot \sum_{|\alpha| \leqslant m-1} \gamma^{2(m-1-|\alpha|)} \| D^\alpha u\|_o^2 \ .$$

Indeed, by (ii)

(1.14) $$(Au,u) = \|Bu\|^2 + (Ru,u) \geqslant (Ru,u) \ .$$

Now if $Q \in OP(S_\gamma^{2\ell})$ then $(\gamma^2-\Delta)^{-\ell} \# Q$ is in $OP(S_\gamma^0)$ and since $(\gamma+|\xi|)^{-|\alpha|} \leqslant (1+|\xi|)^{-|\alpha|}$ for $\gamma \geqslant \gamma_o > 1$ and every α , this operator is L^2 continuous (by the well known continuity theorem for pseudo-differential operators of class $S_{1,0}^0$). So for $u \in C_o^\infty$,

$$|(Qu,u)| = |((\gamma^2-\Delta)^{-\ell} \# Qu,\ (\gamma^2-\Delta)^{\ell} u)| \leqq C\|u\|_o \cdot \|(\gamma^2-\Delta)^{\ell} u\|_o$$

Applying this result to R with $\ell = m-1$ and using (1.14), we get (1.13).

0.3 Pseudo-convexity conditions

Let $P(x,D)$ be a differential operator, defined in a neighborhood of a point $x_o \in \mathbb{R}^n$, with principal symbol $p(x,\xi)$ and S a C^2 hypersurface through x_o,
$S = \{x \in V : \varphi(x) = \varphi(x_o)\}$, $d\varphi(x_o) \neq 0$.

Definition 0.1

The hypersurface S will be called

(i) *pseudo-convex at x_o if for every $\xi \in \mathbb{R}^n$ such that $p(x_o,\xi) = 0$ and $\{p,\varphi\}(x_o,\xi) = 0$, we have $\{p,\{p,\varphi\}\}(x_o,\xi) > 0$.*

(ii) *strongly pseudo-convex at x_o if it is pseudo-convex and moreover :*
For all $\zeta = \xi + i\tau \operatorname{grad}\varphi(x)$, $\xi \in \mathbb{R}^n$, $\tau \neq 0$ satisfying $p(x_o,\zeta) = \{p,\varphi\}(x_o,\zeta) = 0$, we have

$$\sum_{j,k=1}^{n} \frac{\partial^2\varphi}{\partial x_j \partial x_k}(x_o) \frac{\partial p}{\partial \xi_j}(x_o,\zeta) \overline{\frac{\partial p}{\partial \xi_k}(x_o,\zeta)} + \frac{1}{\tau} \operatorname{Im} \sum_{k=1}^{n} \frac{\partial p}{\partial x_k}(x_o,\zeta) \overline{\frac{\partial p}{\partial \xi_k}(x_o,\zeta)} > 0.$$

(iii) *strictly non-pseudo-convex at x_o if there exists $\xi_o \in \mathbb{R}^n \setminus 0$ such that*

$p(x_o,\xi_o) = 0$, $\{p,\varphi\}(x_o,\xi_o) = 0$ *and* $\{p,\{p,\varphi\}\}(x_o,\xi_o) < 0$.

Here $\{f,g\} = \sum_{j=1}^{n} \frac{\partial f}{\partial \xi_j} \cdot \frac{\partial g}{\partial x_j} - \frac{\partial f}{\partial x_j} \cdot \frac{\partial g}{\partial \xi_j}$ denotes the Poisson bracket of f and g.

Remark 0.2.

(i) When $m=1$ it is obvious that there is no difference between pseudo-convexity and strong pseudo-convexity. One can show (see [48]) that the same is true if $m=2$ and p is real.

(ii) Let $(x(s),\xi(s))$ be the null bicharacteristic of p (when it is real) through the point $(x_o,\xi) \in T^* \mathbb{R}^n$. Let us set $\Phi(s) = \varphi(x(s))$. It is easy to see that the condition of pseudo-convexity means that $\Phi(0)=0$, $\Phi'(0)=0$ and $\Phi''(0) > 0$ ■

§1. OPERATORS WITH REAL PRINCIPAL PART

1.1. The uniqueness result.

The aim of this section is to prove the following

Theorem 1.1

Let $P(x,D)$ *be a differential operator of order* m, *defined in a neighborhood* V *of a point* x_o, *with real and* C^2 *principal part and bounded lower order terms. Let* $S = \{x \in V : \varphi(x) = \varphi(x_o)\}$ *be a* C^2 *hypersurface through* x_o *which is strongly pseudo-convex at* x_o. *Then there is a neighborhood* W *of* x_o *such that if* $u \in C^m$ *satisfies* $P(x,D)u = 0$ *and* $u=0$ *in* $\{x \in V : \varphi(x) > \varphi(x_o)\}$ *then* $u=0$ *in* W.

In fact the same conclusion holds for solutions $u \in C^m(V)$ of the differential inequality : $|P(x,D)u(x)| \leq C(u) \sum_{|\alpha|\leq m-1} |D^\alpha u(x)|$.

Let us also remark that S is not supposed to be non-characteristic. Let us give an example. Let V be a neighborhood of the origin in $\mathbb{R}^{n+1} = \{(x,t) : x \in \mathbb{R}^n, t \in \mathbb{R}\}$ and S be the hypersurface $\{(x,t) : t=0\}$. Let P be a second order differential operator of the form

$$P(x,t\ ;\ D_x,D_t) = D_t^2 + a(x,t\ ;\ D_x) + \sum_{j=1}^{n} \alpha_j D_{x_j} + \alpha D_t + \beta$$

where $a(x,t;\xi)$ is a real quadratic form.

Then the condition of strong pseudo-convexity in theorem 1.1, which implies uniqueness, is the following :

"For every $\xi \in \mathbb{R}^n$ such that $a(0,0;\xi)=0$, we have $\frac{\partial a}{\partial t}(0,0;\xi)>0$."

"For instance, we can take $n=2$ and $a(x,t;\xi)=(1+\varepsilon_1 t)\xi_1^2 - (1+\varepsilon_2 t)\xi_2^2$ with $\varepsilon_1 > \varepsilon_2$ ∎

The proof of theorem 1.1 begins as usual with a Carleman estimate.

Theorem 1.2

Under the hypotheses of theorem 1.1, we can find a positive constant C, a neighborhood V of x_o such that for every $\gamma \geqslant e^C$ and all $v \in C_o^\infty(V)$,

$$\sum_{|\alpha|\leqslant m-1} \gamma^{2(m-|\alpha|)-1} \|D^\alpha v\|_o^2 \leqq C\, \|P_\gamma v\|_o^2 \tag{1.15}$$

with $P_\gamma = e^{\gamma\psi}\, P e^{-\gamma\psi}$ and

$$\psi(x)=\varphi(x_o) + \varphi'(x_o)\cdot(x-x_o)+\frac{1}{2}\varphi''(x_o)\cdot(x-x_o)^2 + \frac{C}{2}(\varphi'(x_o).(x-x_o))^2 - \frac{1}{2C^2}\|x-x_o\|^2 .$$

The key to the proof of this result is the following lemma. Let Λ be the p.d.o. with Weyl symbol $(\gamma^2+|\xi|^2)^{-1}$. Let us consider the operator Q_γ defined by

$$Q_\gamma = \frac{C^3}{2}(P_\gamma^\star \Lambda\, P_\gamma + P_\gamma \Lambda P_\gamma^\star) + \frac{1}{2\gamma}[P_\gamma^\star, P_\gamma] \tag{1.16}$$

where C is the same constant as in ψ. Let us denote by $q_C(x,\xi,\gamma)$ the Weyl symbol of Q_γ. Then we have

Lemma 1.3

We can find a constant $C>1$, *such that for* $|x-x_0| \leqslant C^{-2}$ $\gamma \geqslant e^C$ *and* $\xi \in \mathbb{R}^n$,

$$(1.17) \quad q_C(x,\xi,\gamma) \geqslant C^{-1}(\gamma^2+|\xi|^2)^{m-1}.$$

Proof :

By the composition formula (1.10), (1.2) and (1.6), the Weyl symbol of $\frac{C^3}{2}(P^\star_\gamma \wedge P_\gamma + P_\gamma \wedge P^\star_\gamma)$ is equal to $C^3(\gamma^2+|\xi|^2)^{-1}|p(x,\zeta)|^2 + r_{2m-3}(x,\xi,\gamma)$, where r_{2m-3} belongs to S^{2m-3}_γ and $\zeta = \xi + i\gamma\psi'(x)$.

So, by (1.8) and (1.9), the inequality (1.17) will be implied by

$$(1.18) \quad C^3|\zeta|^{-2}|p(x,\zeta)|^2 + C|p'_\xi(x,\zeta)\cdot\varphi'(x_0)|^2 +$$

$$+\varphi''(x_0)\cdot\overline{p'_\xi(x,\zeta)}\cdot p'_\xi(x,\zeta) + \frac{1}{\gamma}\operatorname{Im}\overline{p'_\xi(x,\zeta)}\cdot p'_x(x,\zeta) -$$

$$-\frac{1}{C^2}\overline{p'_\xi(x,\zeta)}\cdot p'_\xi(x,\zeta) + \sum_{j\geqslant 1}\sum_{\ell=0}^{2j+1} a_{2m-2j-\ell-1}(x,\zeta)\,.$$

$$.\gamma^\ell\Pi_\ell(C) \geqslant \frac{\delta o}{C}|\zeta|^{2m-2}$$

where δ_0 is independant of C and depends only on the constants C_0, C_1, C_2 such that :

$$\begin{cases} C_0(|\xi|^2+\gamma^2)^{1/2} \leqslant |\zeta| \leqslant C_1(|\xi|^2+\gamma^2)^{1/2} \\ |r_{2m-3}(x,\xi,\gamma)| \leqslant C_2(|\xi|^2+\gamma^2)^{m-3/2}. \end{cases}$$

To prove (1.18) we use a contradiction argument and say that, otherwise, for every $k \in \mathbb{N}^\star$ we can find x_k, γ_k, ξ_k with $|x_k - x_o| \leqslant k^{-2}$, $\gamma_k \geqslant e^k$, $\xi_k \in \mathbb{R}^n$ such that,

setting

$$Z_k = \frac{\zeta_k}{|\zeta_k|}, \quad \Xi_k = \frac{\xi_k}{|\zeta_k|}, \quad \Gamma_k = \frac{\gamma_k}{|\zeta_k|}$$

and dividing both sides of (1.18) by $|\zeta_k|^{2m-2}$,

(1.19) $$\underset{①}{k^3|p(x_k,Z_k)|^2} + \underset{②}{k|p'_\xi(x_k,Z_k)\cdot\varphi'(x_o)|^2} +$$

$$+ \underset{③}{\varphi''(x_o)\overline{p'_\xi(x_k,Z_k)}\cdot p'_\xi(x_k,Z_k)} + \underset{④}{\frac{1}{\Gamma_k}\operatorname{Im}\overline{p'_\xi(x_k,Z_k)}\cdot p'_x(x_k,Z_k)} -$$

$$- \underset{⑤}{\frac{1}{k^2}\overline{p'_\xi(x_k,Z_k)}\cdot p'_\xi(x_k,Z_k)} + \underset{⑥}{\sum_{j\geqslant 1}\sum_{\ell=0}^{2j+1} a_{2m-2j-\ell-1}(x_k,Z_k)} \cdot$$

$$\cdot \frac{\pi_\ell(k)\cdot\Gamma_k^{2j+\ell-1}}{\gamma_k^{2j-1}} < \frac{\delta_o}{k}.$$

By a compactness argument, we can assume, without loss of generality, that when $k \to +\infty$, $Z_k \to Z_o$, $\Xi_k \to \Xi_o$, $\Gamma_k \to \Gamma_o$.

We now have to consider two different cases, namely $\Gamma_o = 0$ and $\Gamma_o \neq 0$.

<u>Case 1</u> : $\Gamma_o = 0$.

Let us first prove that

(1.20) $$\lim_{k\to\infty} ④ = \sum_{j,\ell=1}^{n} \left(\frac{\partial^2 p}{\partial x_j \partial \xi_\ell}\cdot\frac{\partial p}{\partial \xi_j} - \frac{\partial^2 p}{\partial \xi_j \partial \xi_\ell}\cdot\frac{\partial p}{\partial x_j}\right)(x_o,\Xi_o)\cdot$$
$$\cdot\frac{\partial\varphi}{\partial x_\ell}(x_o).$$

Indeed, let us set

$$F(x,\ \Xi,\ \Gamma,\ c) = \mathrm{Im}\ \overline{p'_\xi(x,\ \Xi + i\Gamma\psi'(x))}.p'_x(x,\ \Xi + i\Gamma\psi'(x))$$

Since p is real, $F(x,\Xi,\ 0,\ c) = 0$ for every (x,Ξ,c).
It follows that

$$\frac{1}{\Gamma}\ F(x,\Xi,\Gamma,c) = \int_0^1 \frac{\partial F}{\partial\Gamma}\ (x,\Xi,t\Gamma)dt\ ,$$

and we get easily

$$\frac{\partial F}{\partial\Gamma} = \mathrm{Re} \sum_{j,\ell} \left(\frac{\partial^2 p}{\partial x_j\ \partial\xi_\ell} \cdot \frac{\partial p}{\partial\xi_j} - \frac{\partial^2 p}{\partial\xi_j\partial\xi_\ell}\ \frac{\partial p}{\partial x_j}\right)\frac{\partial\psi}{\partial x_\ell}\ ,$$

which implies (1.20) since the right hand side of the above formula is a continuous function of (x,Ξ,Γ). In particular, the term ④ is bounded when $k \to \infty$.
Coming back to (1.19), we can write

$$|p(x_k,Z_k)|^2 \leqq \delta_o k^{-1} + k^{-3} \sum_{j=2}^{6} |ⓙ|\ .$$

Obviously $k^{-3}|ⓙ|$ tends to zero for $2 \leqslant j \leqslant 5$. For the term ⑥ we remark that, Γ_k being bounded,

$$\frac{\Pi_\ell(k)}{\gamma_k^{2j-1}} \leqq \frac{k^\ell}{\gamma_k^{2j-1}} \leqq k^2(ke^{-k})^{2j-1} \to 0\ ,\ j \geqslant 1.$$

Now when $k \to \infty$, $Z_k = \Xi_k + i\Gamma_k\psi'(x_k)$ tends to Ξ_o. So we have proved

$$(1.21) \qquad p(x_o,\ \Xi_o) = 0\ .$$

Let us remark that since $\Gamma_k \to 0$, we have $\dfrac{|\xi_k|}{\gamma_k} \to \infty$, so

$\Xi_o \neq 0$.

Again by (1.19),

$$|p'_\xi(x_k,Z_k).\varphi'(x_o)|^2 \leqq \delta_o k^{-1} + k^{-1} \sum_{j=3}^{6} |ⓙ|\ ,$$

and by the same argument as before,

$$(1.22)\quad \{p,\varphi\}(x_o, \Xi_o) = \sum_{j=1}^{n} \frac{\partial p}{\partial \xi_j}(x_o, \Xi_o)\frac{\partial \varphi}{\partial x_j}(x_o) = 0 .$$

Using one more time (1.19), we get

$$③ + ④ \leqq \delta_o k^{-1} + |⑤| + |⑥| .$$

Now the right hand side of this inequality tends to zero, so it follows from (1.20) that

$$(1.23)\qquad \lim_{k\to\infty} [③ + ④] = \{p,\{p,\varphi\}\}(x_o,\Xi_o) \leqslant 0 .$$

This ends the proof of the Lemma, in the Case 1, since (1.21), (1.22) and (1.23) are in contradiction with the pseudo-convexity of the hypersurface S .

<u>Case 2</u> : $\Gamma_o \neq 0$.

In this case $\lim\limits_{k\to\infty} Z_k = Z_o = \Xi_o + i\Gamma_o\,\varphi'(x_o)$. Coming back to (1.19) we see that

$$(1.24)\qquad \lim_{k\to\infty} ④ = \frac{1}{\Gamma_o}\,\mathrm{Im}\,\overline{p'_\xi(x_o,Z_o)}.p'_x(x_o,Z_o) ,$$

Using exactly the same argument as before, we see that

$$p(x_o,Z_o) = \{p,\varphi\}(x_o,Z_o) = 0$$

and

$$\varphi''(x_o).\overline{p'_\xi(x_o,Z_o)}.p'_\xi(x_o,Z_o) + \frac{1}{\Gamma o}\,\mathrm{Im}\,\overline{p'_\xi(x_o,Z_o)}.p'_x(x_o,Z_o) \leqslant 0 ,$$

which is again in contradiction with the strong pseudo-convexity of S and gives a complete proof of Lemma 1.3 ∎

<u>Proof of theorem 1.2</u>

Let the constant C be fixed by the Lemma 1.3. Then the operator $A = \frac{C^3}{2}(P^*_\gamma \Lambda P_\gamma + P_\gamma \Lambda P^*_\gamma) + \frac{1}{2\gamma}[P^*_\gamma, P_\gamma] - \frac{1}{2C}(\gamma^2 - \Delta)^{m-1}$

is in S_γ^{2m-2} and has, by Lemma 1.3, a symbol $a(x,\xi,\gamma)$ which satisfies

$$a(x,\xi,\gamma) \geq \frac{1}{2C}(\gamma^2 + |\xi|^2)^{m-1} .$$

So, by (1.13) we can find a constant $M > 0$ such that for $u \in C_o^\infty(V_{x_o})$,

$$(1.25) \qquad (Au,u) \geq -M \sum_{|\alpha| \leq m-1} \gamma^{2(m-2-|\alpha|)} \| D^\alpha u \|_o^2 .$$

$$\text{Now } (Au,u) = \frac{C^3}{2}(\Lambda P_\gamma u, P_\gamma u) + \frac{C^3}{2}(\Lambda P_\gamma^* u, P_\gamma^* u) + \frac{1}{2\gamma}\|P_\gamma u\|^2 - \frac{1}{2\gamma}\|P_\gamma^* u\|^2 -$$

$$- \frac{1}{2C}((\gamma^2 - \Delta)^{m-1} u,u) .$$

Since we have $(\Lambda W,W) \leqq \frac{1}{\gamma^2}\|W\|_o^2$, $W \in C_o^\infty$ and $((-\Delta)^j u,u) + \|u\|_o^2 \geqq \|u\|_j^2$, we deduce from (1.25) that, taking γ big enough, $(\frac{1}{2\gamma} + \frac{C^3}{\gamma^2})\|P_\gamma u\|_o^2 - (\frac{1}{2\gamma} - \frac{C^3}{\gamma^2})\|P_\gamma^* u\|_o^2 \geq$

$$\geq \delta_o \sum_{|\alpha| \leq m-1} \gamma^{2(m-1-|\alpha|)} \|D^\alpha u\|_o^2 .$$

Taking γ such that $\frac{1}{2\gamma} - \frac{C^3}{\gamma^2} > 0$, we get (1.15). The proof of theorem 1.2 is complete. ∎

Proof of theorem 1.1

Let $u \in C^m(V)$ be a solution of $Pu = 0$ with support in $\{\varphi(x) \leq \varphi(x_o)\}$ and $\chi \in C_o^\infty(V)$, $\chi = 1$ on $V_1 \subset\subset V$.

By the inequality (1.15) of theorem 1.2

$$(*) \qquad \|e^{\gamma\psi}\chi u\|^2 \leq C\|e^{\gamma\psi}P\chi u\|_o^2 = C\|e^{\gamma\psi}[P,\chi]u\|_o^2 .$$

We shall prove that

$$(**) \qquad \varphi(x) \leq \varphi(x_o) \text{ implies } \psi(x) < \psi(x_o) \text{ if } x \in V_{x_o} , x \neq x_o .$$

Assuming (**), on the support of $[P,\chi]u$ we have $\psi(x) \leq \psi(x_o)-\varepsilon_o$ $\varepsilon_o>0$, since $x_o \notin \operatorname{supp}[P,\chi]u$. Let us set $W = \{x : \psi(x) \geq \psi(x_o) + \frac{\varepsilon_o}{2}\}$. It is a neighborhood of x_o. Let V_2 be a neighborhood of x_o contained in $V_1 \cap W$ and $\chi_1 \in C_o^\infty(V_2)$, $\chi_1 = 1$ on $V_3 \subset\subset V_2$. By (*) we have

$$e^{\gamma(\psi(x_o)+\frac{\varepsilon_o}{2})} \|\chi_1 u\|^2_{L^2(V_2)} \leqq \|e^{\gamma\psi}\chi_1 u\|^2_{L^2(V_1)} \leq C_o \|e^{\gamma\psi}u\|^2_{L^2(V_1)}$$

$$= C_o \|e^{\gamma\psi}\chi u\|^2_{L^2(V_1)} \leq C_o C\|e^{\gamma\psi}[P,\chi]u\|^2_{L^2(V)} \leq C_o C\, e^{\gamma(\psi(x_o)-\varepsilon_o)}$$

$$\cdot\|[P,\chi]u\|^2_{L^2(V)},$$

so

$$\|\chi_1 u\|^2_{L^2(V_2)} \leqq C_1 e^{-\frac{1}{2}\gamma\varepsilon_o} \|[P,\chi]u\|^2_{L^2(V)} .$$

Letting $\gamma \to +\infty$ we get $u \equiv 0$ on V_3.

Let us now prove (**). We recall that

$$\psi(x) = \varphi(x_o) + \varphi'(x_o).(x-x_o) + \frac{1}{2}\varphi''(x_o).(x-x_o)^2 +$$

$$+ \frac{C}{2}[\varphi'(x_o).(x-x_o)]^2 - \frac{1}{2C^2}\|x-x_o\|^2,$$

It follows that

$$\psi(x)-\psi(x_o) = [\varphi'(x_o).(x-x_o) + \frac{1}{2}\varphi''(x_o).(x-x_o)^2](1+\frac{C}{2}\varphi'(x_o).(x-x_o)) -$$

$$- \frac{C^2}{4}\varphi'(x_o).(x-x_o)\varphi''(x_o).(x-x_o)^2 - \frac{1}{2C^2}\|x-x_o\|^2.$$

Writing $\varphi'(x_o).(x-x_o) + \frac{1}{2}\varphi''(x_o).(x-x_o)^2 = \varphi(x)-\varphi(x_o) +$

$$+ \lambda(x)(x-x_o),$$

we get

$$\psi(x)-\psi(x_o) = (\varphi(x)-\varphi(x_o))(1+\frac{C}{2}\varphi'(x_o).(x-x_o))-\|x-x_o\|^2F(x)$$

where $F(x_o) = \frac{1}{2C^2}$, from which we deduce (**) if V_{x_o} is small enough. The proof of Theorem 1.1 is complete. ■

Let us come now to the necessity of the pseudo - convexity condition.

1.2 The non-uniqueness result.

The main result of this section is the following

Theorem 1.4

Let P be a differential operator of order m with C^∞ coefficients, in a neighborhood V of a point x_o, with real principal part. Let $S=\{x\in V : \varphi(x) = \varphi(x_o)\}$ be a non-characteristic C^∞ hypersurface through x_o which is strictly non-pseudo-convex at x_o; i.e.

(i) $\exists\ \xi_o\in\mathbb{R}^n\setminus o : p(x_o,\xi_o) = \{p,\varphi\}(x_o,\xi_o) = 0$ *and*

$$\{p,\{p,\varphi\}\}(x_o,\xi_o) < 0 .$$

Let us suppose, moreover, that

(ii) $\{\{p,\varphi\},\varphi\}(x_o,\xi_o) \neq 0$

(iii) $dp(x_o,\xi_o)\neq 0$ *and the manifold* $\Sigma=\{(y,\eta)\in T^*\mathbb{R}^n : p(y,\eta) = 0\}$ *is transverse to the vertical fibers.*

Then there exist a neighborhood W of x_o and two $C^\infty(W)$ functions u and a, vanishing in $\{x\in W : \varphi(x) > \varphi(x_o)\}$, such that

$$\begin{cases} Pu + au = 0 \\ x_o \in \operatorname{supp} u . \end{cases}$$

Remarks 1.5.

a) The necessity of condition (iii) is shown by the Tricomi operator $P = D_t^2 + tD_x^2$ with $\xi_o = (1,0)$, for which $P + a$ possess the (two-sided) uniqueness property with respect to $S = \{t=0\}$.

b) With the notations of the example after theorem 1.1, let P be the operator in $\mathbb{R}^{n+1}$: $P = D_t^2 + a(x,t\,;D_x)$. Then Theorem 1.4 applies if

1) $\exists\ \xi_o \in \mathbb{R}^n\backslash 0 : a(0,0;\xi_o) = 0$ and

$\frac{\partial a}{\partial t}(0,0\,;\xi_o) < 0$

2) $\text{grad}_\xi\, a(0,0\,;\xi_o) \neq 0$.

For example $n = 2$, $a(x,t\,;\xi) = (1+\varepsilon_1 t)\xi_1^2 - (1+\varepsilon_2 t)\xi_2^2$ with $\varepsilon_1 < \varepsilon_2$ and $\xi_o = (1,1)$.

Proof of Theorem 1.4

All the ingredients being invariant under a smooth change of coordinates, we can assume that we work in the $(x,t) \in \mathbb{R}^{n-1} \times \mathbb{R}$ coodinates, near the origin, where S is given by $\{(x,t) : t = 0\}$ and $P = P(x,t\,;D_x, D_t)$. As in the proof of theorem 3.1 Chapter One, let δ be a positive number, $\lambda = \delta^{-\theta}$ with $\theta > 1$ and $s = \lambda(t-\delta)$. Our operator becomes $P(x,\delta + \frac{s}{\lambda}\,;D_x, \lambda D_s)$.
The approximate solution will be taken in the form

$$(1.26)\quad u = e^{i\sigma(\tilde{\xi}(x,\delta) + \frac{s}{\lambda}\tau(x,\delta))}\, e^{\nu\varphi(x,\delta,s)}\, e^{-\gamma(x,\delta)}\, w(x,\delta,s).$$

Let us define $\tilde{P}$ by $\frac{Pu}{u} = \frac{\tilde{P}w}{w}$. Then we have

$$(1.27)\quad \tilde{P} = P(x,\delta + \frac{s}{\lambda}\,;\sigma\nabla_x\tilde{\xi} + \sigma\frac{s}{\lambda}\nabla_x\tau + \nu\nabla_x\varphi - \nabla_x\gamma + D_x, \sigma\tau D_s\varphi + {} + \lambda D_s).$$

A.Choice of $\tilde{\xi}$ and τ

Lemma 1.6

Let p be the principal symbol of P and $\eta^\circ = (\xi^\circ, \tau^\circ) \in \mathbb{R}^n \setminus 0$ satisfy the conditions of theorem 1.4. Then we can find two real functions $\xi(x,\delta)$, $\tau(x,\delta)$, smooth for $\delta \geqslant 0$ and x near the origin, such that

$$(1.28)\quad \begin{cases} p(x,\delta;\ \nabla_x\xi(x,\delta),\ \tau(x,\delta)) = \dfrac{\partial p}{\partial \tau}(x,\delta;\ \nabla_x\xi(x,\delta),\tau(x,\delta))=0 \\ (\nabla_x\xi(0,0),\ \tau(0,0)) = \eta^\circ \end{cases}$$

and

$$(1.29)\qquad \{p,\ \{p,t\}\}(\eta^\circ) = -\Big[\frac{\partial^2 p}{\partial \tau^2}\Big(\frac{\partial p}{\partial t} + \sum_{j=1}^{n-1} \frac{\partial p}{\partial \xi_j}\cdot \frac{\partial \tau}{\partial x_j}\Big)\Big](\eta^\circ).$$

Proof.

By condition (ii) of Theorem 1.4 we have $\dfrac{\partial^2 p}{\partial \tau^2}(0,0\ ;\ \xi_o,\tau_o) \neq 0$, so the equation $\dfrac{\partial p}{\partial \tau}(x,t\ ;\xi,\tau) = 0$ is equivalent near $(0,0\ ;\ \xi_o,\tau_o)$ to $\tau = q(x,t\ ;\xi)$ for some real function q and we have $q(0,0\ ;\ \xi_o) = \tau_o$.
Let us set $F(x,t\ ;\xi) = p(x,t\ ;\xi,\ q(x,t\ ;\xi))$. It is easy to see that, by (iii), $\dfrac{\partial F}{\partial \xi_j}(0,0\ ;\xi_o) = \dfrac{\partial p}{\partial \xi_j}(0,0\ ;\xi_o) \neq 0$, so by the Hamilton-Jacobi theory the problem

$$\begin{cases} F(x,\ \delta\ ;\ \nabla_x\xi(x,\delta)) = 0 \\ \nabla_x\xi(0,0) = \xi_o \end{cases}$$

admits a unique smooth solution $\xi(x,\delta)$ for $\delta \geqslant 0$ and x near zero. Setting $\tau(x,\delta) = q(x,\delta\ ;\nabla_x\xi(x,\delta))$, it is easily seen that we get (1.28). Now the equations of the bicharacteristic through η° are

$$\dot{x} = p'_\xi\ ,\ \dot{t} = p'_\tau\ ,\ \dot{\xi} = -p'_x\ ,\ \dot{\tau} = -p'_t\ ,$$

$$(x(0),\ t(0),\ \xi(0),\ \tau(0)) = \eta^\circ.$$

It follows that

$$\ddot{t} = p''_{\tau x} \cdot \dot{x} + p''_{\tau t} \cdot \dot{t} + p''_{\tau\xi} \cdot \dot{\xi} + p''_{\tau\tau} \cdot \dot{\tau} = p''_{\tau x} \cdot p'_{\xi} + p''_{\tau t} \cdot p'_{\tau} -$$

$$- p''_{\tau\xi} \cdot p'_{x} - p''_{\tau\tau} \cdot p'_{t} \,.$$

Taking the derivative of (1.28) with respect to x, we get

$$\begin{cases} p'_{x} + \xi''_{xx} \cdot p'_{\xi} + \tau'_{x} \cdot p'_{\tau} = 0 \\ p''_{\tau x} + \xi''_{xx} \cdot p''_{\tau\xi} + \tau'_{x} \cdot p''_{\tau\tau} = 0 \,. \end{cases}$$

At the point η_0 we have the equality

$$\ddot{t} = -(\xi''_{xx} \cdot p''_{\tau\xi}) \cdot p'_{\xi} - p''_{\tau\tau}\, \tau'_{x} \cdot p'_{\xi} + (\xi''_{xx} \cdot p'_{\xi}) p''_{\tau\xi} - p''_{\tau\tau} \cdot p'_{t}$$

from which we deduce (1.29). ■

We shall take in (1.26) $\tilde{\xi}(x,\delta) = \xi(x,\delta) + \frac{1}{\lambda} C.x$ where C is a constant vector in $\mathbb{R}^{n-1}$ which will be choosen later and $\xi(x,\delta)$, $\tau(x,\delta)$ given by the above Lemma.

B. Choice of the phase φ

Lemma 1.7

There exists $C \in \mathbb{R}^{n-1}$ and $\varphi(x,\delta,s)$ smooth in $V_0 \times [0,\delta_0[\times]-s_0, s_0[$ such that with $\tilde{\xi}(x,\delta) = \xi(x,\delta) + \frac{1}{\lambda} C.x$ and $\sigma = \lambda^{3/2}.\nu$, we have

(1.30) $$p(x,\delta + \frac{s}{\lambda} \,;\, \nabla_x \tilde{\xi} + \frac{s}{\lambda} \nabla_x \tau,\ \tau + \frac{\lambda\nu}{\sigma} D_s \varphi) = 0 .$$

Moreover

(1.31) $$\operatorname{Re} \varphi(x,\delta,s) = \alpha(x,\delta) \cdot s - \beta(x,\delta,s) \cdot s^2$$

with α and β smooth and

(1.32) $$\alpha(0,0) > 0 \ ,\ \beta(0,0,0) > 0 \ .$$

Proof :

Let us set $G(x,\delta,s,Z) = \lambda p(x,\delta+s\delta^{\theta} ; \nabla\tilde{\xi} + s\delta^{\theta}\nabla_x\tau, \tau + \delta^{\frac{\theta}{2}}Z)$ (we recall that $\lambda = \delta^{-\theta}$ and $\frac{\lambda\nu}{\sigma} = \delta^{\theta/2}$).

We first find a Z^o such that $G(0,0,0,Z^o)=0$ and $\frac{\partial G}{\partial Z}(0,0,0;Z_o) \neq 0$. Then, using the implicit function theorem we shall find a solution $Z = Z(x,\delta,s)$ of

$$(1.33)\quad \begin{cases} p(x,\delta+s\delta^{\theta} ; \nabla\tilde{\xi}+s\delta^{\theta}\nabla_x\tau,\tau+\delta^{\theta/2}z) = 0 \\ Z(0,0,0) = Z^o. \end{cases}$$

Setting $X= (x,\delta+s\delta^{\theta} ;\nabla\tilde{\xi} + s\delta^{\theta}\nabla_x\tau , \tau)$ we can write

$$G = G(x,\delta,s,Z) = p(X) + \delta^{\theta/2} Z \frac{\partial p}{\partial\tau}(X) + \frac{1}{2}\delta^{\theta} Z^2 \frac{\partial^2 p}{\partial\tau^2}(X) + 0(\delta^{\frac{3\theta}{2}}).$$

$$G = p(x,\delta + s\delta^{\theta} ; \nabla_x\xi,\tau) + p'_{\xi}(x,\delta + s\delta^{\theta} ; \nabla_x\xi,\tau)\cdot$$

$$\cdot(\frac{C}{\lambda} + \frac{s}{\lambda}\nabla_x\tau) + 0(\delta^{2\theta}) + \frac{\partial p}{\partial\tau}(x,\delta+s\delta^{\theta};\nabla_x\xi,\tau)\delta^{\theta/2}Z + 0(\delta^{3\theta/2})+$$

$$+ \frac{1}{2}\cdot\frac{\partial^2 p}{\partial\tau^2}(x,\delta + s\delta^{\theta} ; \nabla_x\xi ,\tau)\delta^{\theta} Z^2 + 0(\delta^{2\theta}) + 0(\delta^{3\theta/2}).$$

Using once more the Taylor formula and (1.28), we get

$$(1.34)\quad \begin{aligned} \lambda G = {} & s[p'_t(x,\delta;\nabla_x\xi,\tau) + p'_{\xi}(x,\delta;\nabla_x\xi,\tau)\cdot\tau'_x] + C.p'_{\xi}(x,\delta;\nabla_x\xi,\tau) \\ & + \frac{1}{2}p''_{\tau\tau}(x,\delta;\nabla_x\xi,\tau)Z^2 + 0(\delta^{\theta/2}). \end{aligned}$$

Let us set

$$A = C\cdot p'_{\xi}(0,0;\xi_o,\tau_o),\ B = p'_t(0,0;\xi_o,\tau_o)+p'_{\xi}(0,0;\xi_o,\tau_o)\cdot\tau'_x(0,0).$$

Since $p'_{\xi}(0,0;\xi_o,\tau_o) \neq 0$ we can find C such that $A=-B$.

On the other hand, by (1.29) and by condition (i) of theorem 1.4 ,

$$\frac{B}{p''_{\tau\tau}(0,0;\xi_o,\tau_o)} = \frac{-\ddot{t}(0)}{[p''_{\tau\tau}(0,0;\xi_o,\tau_o)]^2} < 0 .$$

Finally, we take as Z^o the square root of $\dfrac{2B}{p''_{\tau\tau}(0,0;\xi_o,\tau_o)}$

which has negative imaginary part.

It is then easy to see that $\frac{\partial G}{\partial Z}(0,0,0,Z_o) \neq 0$.

Let us define φ by $\frac{1}{i}\frac{\partial\varphi}{\partial s} = Z(x,\delta,s)$, (Z given by (1.33))

with $\varphi|_{s=0}=0$. We have $\frac{\partial^2\varphi}{\partial s^2} = iZ'_s$. Now taking the derivative, with respect to s, of the equation (1.30) and taking $x = \delta = s = 0$, we get, with above notation ,

$$B + p''_{\tau\tau}(0,0;\xi_o,\tau_o)\cdot Z_o\cdot Z'_s(0,0,0)=0.$$

It follows, since $Z_o^2 = \dfrac{2B}{p''_{\tau\tau}(0,0;\xi_o,\tau_o)}$, that

$B\,Z_o + 2B\,Z'_s(0,0,0) = 0$, so $Z'_s(0,0,0) = -\frac{1}{2}Z_o$ and $\mathrm{Re}\,\varphi''(0,0,0) = -\mathrm{Im}\,Z'_s(0,0,0) = \frac{1}{2}\mathrm{Im}\,Z_o < 0$, which proves (1.31) and (1.32)∎

<u>Corollary 1.8</u>

<u>With the notations of Lemma</u> 1.7,

$$P(x,\delta+\frac{s}{\lambda};\nabla_x\tilde{\xi}+\frac{s}{\lambda}\nabla_x\tau,\tau+\frac{\lambda\nu}{\sigma}(D_s\varphi+\frac{1}{\nu}D_s))=$$

(1.35)

$$=\frac{1}{\lambda\nu}\{H(s,x)D_s + K(\delta,x) + \delta^r Q(x,\delta,s;D_s,D_x)\}$$

<u>where</u> H,K <u>and the coefficients of</u> Q <u>are regular</u>,

$H(0,0) \neq 0$ <u>and</u> $r > 0$.

<u>Proof</u> :

$$P(x,\delta + \frac{s}{\lambda} ; \nabla_x \tilde{\xi} + \frac{s}{\lambda} \nabla_x \tau , \tau + \delta^{\theta/2} (D_s \varphi + \frac{1}{\nu} D_s)) =$$

$$= \sum_{|\alpha|+k=m} a_{\alpha\beta}(x,\delta+s\delta^\theta) (\nabla_x \tilde{\xi} + s\delta^\theta \nabla_x \tau)^\alpha (\tau + \delta^{\theta/2} (D_s \varphi + \frac{1}{\nu} D_s))^k.$$

We see easily that

$$(\tau + \delta^{\theta/2} (D_s \varphi + \frac{1}{\nu} D_s))^k = (\tau + \delta^{\theta/2} D_s \varphi)^k +$$

$$+ k(\tau + \delta^{\theta/2} D_s \varphi)^{k-1} \frac{\delta^{\theta/2}}{\nu} D_s - a_k \tau^{k-2} \frac{\delta^\theta}{\nu} \varphi''_{ss} + \delta^r \frac{\delta^\theta}{\nu} R(x,\delta,s;D_s)$$

where a_k is a regular function of (x,δ,s), $r > 0$ and R is a differential operator with regular coefficients.

It follows, using Lemma 1.7, that

$$p(x,\delta + s\delta^0 ; \nabla_x \tilde{\xi} + s\delta^\theta \nabla_x \tau , \tau + \delta^{\theta/2} (D_s \varphi + \frac{1}{\nu} D_s)) =$$

$$= \frac{\partial p}{\partial \tau} (x,\delta + s\delta^\theta ; \nabla_x \tilde{\xi} + s\delta^\theta \nabla_x \tau , \tau + \delta^{\theta/2} D_s \varphi) \frac{\delta^{\theta/2}}{\nu} D_s +$$

$$+ \frac{\delta^\theta}{\nu} K(x,s,\delta) + \delta^r \frac{\delta^\theta}{\nu} Q(x,\delta,s;D_x,D_s).$$

Applying Taylor's formula to $\frac{\partial p}{\partial \tau}(\dots)$, we see easily that

$$p(\dots) = \frac{\partial^2 p}{\partial \tau^2} (x,\delta;\nabla_x \xi,\tau) (D_s \varphi) \frac{\delta^\theta}{\nu} D_s + \frac{\delta^\theta}{\nu} K + \delta^r \frac{\delta^\theta}{\nu} Q$$

$$= \frac{1}{\lambda\nu} [p''_{\tau\tau} (x,\delta;\nabla_x \xi,\tau) \cdot (D_s \varphi) \cdot D_s + K(x,\delta,s) +$$

$$+ \delta^r Q(x,\delta,s;D_x,D_s)]$$

with $r > 0$. The proof is complete ■

C. Fixing the parameters

As usual all the parameters $\delta,\lambda,\nu,\sigma$ ... will depend on an integer k. Our choice will be the following

$$\theta = \frac{3}{2}\ ,\ \delta_k = k^{-\rho}\ ,\ \rho = \frac{9}{5}\ ,\ \nu_k = k^{\varepsilon},\ \varepsilon = \frac{1}{3}\ ,\ \lambda_k = \delta_k^{-\theta}\ ,$$

$$\sigma_k = \lambda_k^{3/2}\ \nu_k\ ,\quad \ell_k = \delta_k - \delta_{k+1}.$$

However for sake of simplicity we will keep the notation $\delta,\lambda,\nu,\sigma\ldots$ ∎

We shall use the function

$$(1.36)\quad u_k(x,t) = e^{-\gamma(\delta_k,x)}\, e^{i\sigma_k(\tilde{\xi}(\delta_k,x) + \tau(\delta_k,x)(t-\delta_k))}\ .$$

$$.\ e^{\nu_k\varphi(x,\delta_k,s)}\ .\ W(x,\delta_k,s)$$

for $t \in [\delta_{k+1}\ ,\ \delta_{k-1}]$. Then

$$|s| = \lambda_k\ |t-\delta_k| \leqslant \lambda_k(\delta_{k-1} - \delta_{k+1}) \sim k^{\rho\theta}\cdot\rho k^{-\rho-1}\ ,$$

and since $\rho(\theta-1)-1 = -\frac{1}{10} < 0$, $|s|$ is near the origin and φ is well defined.

D. Choosing the function γ

Let us set

$$(1.37)\quad G_k(x,t) = \sigma_{k+1}\ \mathrm{Im}\ \tau(x,\delta_{k+1})(t-\delta_{k+1}) -$$

$$- \sigma_k\ \mathrm{Im}\ \tau(x,\delta_k)(t-\delta_k) + \nu_k\ \mathrm{Re}\ \varphi(x,\delta_k,s_k) -$$

$$- \nu_{k+1}\ \mathrm{Re}\ \varphi\ (x,\delta_{k+1},s_{k+1})$$

where $s_k = \lambda_k(t-\delta_k)$.

Lemma 1.9

Let us set $m_k = \frac{1}{3}\delta_k + \frac{2}{3}\delta_{k+1}$ *and* $I_k(x) = G_k(x, m_k)$.

Then

$$(1.38) \qquad I_k(x) \sim -\alpha(x,0)\,\rho k^{\rho(\theta-1)+\varepsilon-1}, \quad k \to +\infty .$$

Proof :

We use (1.31) and the method of Lemma 3.3 Chap. 1. ▪

Now, we set

$$(1.39) \qquad \gamma_k(x) = -\sum_{j=k_o}^{k-1} I_j(x).$$

Then $\gamma_{k+1}(x) - \gamma_k(x) = -I_k(x)$. We choose a regular function $\gamma(x,\delta)$ such that $\gamma(x,\delta_k) = \gamma_k(x)$. It follows that

$$(1.40) \qquad \gamma(x,\delta) = \alpha(x,0)\,\frac{\rho}{\rho(\theta-1)+\varepsilon}\,\delta^{-d}, \quad d = \theta - 1 + \frac{\varepsilon}{\rho}.$$

E . The transport equations

Let us recall that the operator $\tilde{P}$ defined by (1.27) has the form

$$\tilde{P} = \sigma^m P_m(x, \delta + s\delta^\theta; \nabla_x\xi + \frac{s}{\lambda}\nabla_x\tau + \frac{\nu}{\sigma}\nabla_x\varphi - \frac{1}{\sigma}\nabla_x\gamma + \frac{1}{\sigma}D_x, \tau + \frac{\lambda\nu}{\sigma}D_s\varphi + \frac{\lambda}{\sigma}D_s) +$$

$$+ \sigma^{m-1}P_{m-1}(\dots) + \dots$$

Lemma 1.10

$$(1.41) \qquad \sigma^{-m}\tilde{P} = \frac{1}{\lambda\nu}\,[H(x,\delta)\,D_s + K(x,\delta) + \delta^{\tilde{r}}Q(x,\delta,s;D_x,D_s)]$$

where H, K *and the coefficients of* Q *are regular*, $H(0,0) \neq 0$, $\tilde{r} > 0$.

Proof:

The terms of $\sigma^{-m}\tilde{P}$ coming from the lower order terms will be included in the term $\frac{1}{\delta\nu}\delta^{\tilde{r}}Q$. Concerning the principal term P_m, we see first that the term $\frac{1}{\sigma}\nabla_x\gamma$ is $o(\frac{1}{\lambda\nu})$ since $\frac{\delta^{-d}}{\sigma} = \frac{\delta^{-d}}{\lambda^{3/2}\nu} = \frac{1}{\lambda\nu} \cdot \frac{\delta^{-d}}{\lambda^{1/2}} = \frac{1}{\lambda\nu}\, \delta^{-d+\frac{1}{2}\theta}$ and $-d + \frac{1}{2}\theta = 1 - \frac{1}{2}\theta - \frac{\varepsilon}{\rho} > 0$. In the same way, the term $\frac{\nu}{\sigma}\nabla_x\varphi$ is $o(\frac{1}{\lambda\nu})$. So, using the Taylor formula and the corollary 1.10, we get (1.41). ∎

We can now solve the Cauchy problems

$$HD_s W_o + KW_o = 0 \qquad HD_s W_j + KW_j = -QW_{j-1} \quad , \quad j \geq 1$$

$$W_o(x,\delta,0) = 1 \qquad W_j(x,\delta,0) = 0$$

and patch together the W_j to get a formal solution $\hat{W} = \sum_{j \geq 0} W_j\, \delta^{j\tilde{r}}$ of $(HD_s + k + \delta^{\tilde{r}}Q)\hat{W} = 0$.

Now we take a regular function $Z(x,\delta,s,\zeta)$ such that

$$|D_x^\alpha D_s^\ell (Z - \sum_{j=0}^{N} W_j\zeta^j)| \leq C_{\alpha\ell N} |\zeta|^{N+1}$$

and we set $W(x,\delta,s) = Z(x,\delta,s,\delta^{\tilde{r}})$. We get a regular function **W** which satisfies $(HD_\delta + K + \delta^{\tilde{r}}Q)W = 0(\delta^\infty)$. ∎

Lemma 1.11

Let us set

(1.42) $h_k(x,t) = e^{-\gamma(x,\delta_k)}\, e^{i\sigma_k(\tilde{\xi}(x,\delta_k) + i\tau(x,\delta_k)(t-\delta_k))}\,.$

$$\cdot\, e^{\nu_k\varphi(x,\delta_k,s)}\, W(x,\delta_k,s)$$

where $s = \lambda_k(t-\delta_k)$, *and* $r_k = \frac{Ph_k}{h_k}$, $t \in [\delta_{k+1}, \delta_{k-1}]$, $x \in V$.

Then for every $\alpha \in \mathbb{N}^{n+1}$, $N \in \mathbb{N}$, *we can find* $C_{\alpha,N} > 0$

such that

$$|D^{\alpha} r_k(x,t)| \leqq C_{\alpha,N} k^{-N}$$

for $k \geqslant k_o$ *and* $(x,t) \in V \times [\delta_{k+1}, \delta_{k-1}]$.

Proof : See Prop. 3.4 Chap 1. ■

F. The set $|h_k| = |h_{k+1}|$

Lemma 1.12

There exists a positive constant C *such that for* $(x,t) \in V \times [\delta_{k+1}, \delta_{k-1}]$,

(1.43) $$\left|\frac{\partial F_k}{\partial t}(x,t)\right| \geqq C\, k^{\eta}$$

where $F_k = \mathrm{Log}\left|\frac{h_k}{h_{k+1}}\right|$ *and* $\eta = 2\rho\theta + \varepsilon - \rho - 1 = \frac{44}{15} > \rho + 1 = \frac{14}{5}$.

Proof : See Prop. 3.5 Chap 1. ■

The end of the proof is made along the same lines as the proof of Theorem 3.1 Chap. 1, so we do not repeat the arguments and refer the reader to that proof. ■

§2. A CLASS OF QUASI-HOMOGENEOUS OPERATORS

Let Ω be an open set of $\mathbb{R}^{n+1} = \{(x,y) : x \in \mathbb{R}^n,\ y \in \mathbb{R}\}$. We consider in this section the following differential operator in a neighborhood V of a point $(x_o, y_o) \in \Omega$:

(2.1) $$P(x,y\,;D_x,D_y) = \sum_{|\alpha| \leqslant 2} a_{\alpha}(x,y)\, D_x^{\alpha} + c(x,y)\, D_y .$$

We shall denote by $p(x,y\,;\xi,\eta)$ the principal symbol of P and assume that

(2.2) $$\begin{cases} p(x,y\,;\xi,\eta) \text{ and } c(x,y) \text{ are real and } C^{\infty} \\ c(x_o,y_o) \neq 0. \end{cases}$$

In order to state the uniqueness and non-uniqueness results, we have to define what we call "pseudo-convex" hypersurfaces. So we begin with some geometrical considerations.
Let σ be the symplectic two form on $T^*\Omega$ and recall that a submanifold N of $T^*\Omega$ is called involutive if for every $m \in N$, $T_m N^{\perp} \subset T_m N$, where the orthogonality is defined by σ.
It is easy to see that if N is given by

$$N = \{(x,y\,;\xi,\eta) \,:\, u_1(x,y\,;\xi,\eta) = \ldots = u_k(x,y\,;\xi,\eta) = 0\}$$

with du_i independant on N, then N involutive is equivalent to say that $\{u_i,u_j\} = 0$ on N for every $i,j=1,\ldots,k$.
Indeed, we see easily that we have

$$T_m N = \{H_{u_1}(m),\ldots,H_{u_k}(m)\}^{\perp} \ \forall\, m \in N\,,$$

which means that the tangent plane to N at m is equal to the orthogonal (with respect to σ) of the space generated by $H_{u_1}(m),\ldots,H_{u_k}(m)$, where $H_{u_i}(m)$ denotes the Hamiltonian of u_j taken at m.

Since $[H_{u_i},H_{u_j}] = H_{\{u_i,u_j\}} = 0$, we deduce from the Frobenius theorem that the tangent system $m \to \{H_{u_1}(m),\ldots,H_{u_k}(m)\} = T_m N^{\perp} \subset T_m N$ is integrable, which means that for every $m_o \in N$ there exists a submanifold F containing m_o such that for every $m \in F$, $T_m F = T_m N$. We shall take in the following

$$N = \{(x,y,\xi,\eta) \in T^*\Omega\backslash 0 \,:\, \xi = 0\}.$$

It is then easy to see that

$$F = \{(x,y,\xi,\eta) \in T^*\Omega\setminus 0 : y=y_o,\ \eta=\eta_o,\ \xi=0,\ x \in \mathbb{R}^n\}.$$

Let us note that $N \subset p^{-1}(0)$ and that p has double characteristics on N.
Let Q^F be the quadratic form induced by $\frac{1}{2}$ Hess p on $T_F(M)$

(where Hess p denotes the Hessian of p). We are going, using Q^F, to define an invariant on T^*F. Let us denote $M = T^*\Omega$. We note first that σ induces an isomorphism $i : TM \to T^*M$. Indeed, for $(z,t) \in TM$ we set $i(z,t) = (z, i_z(t))$ where $i_z(t).s = \sigma_z(t,s)$.

This isomorphism i induces another one, $i_F : T_F M \to T_F^* M$.

Now, obviously, we have a surjective morphism $\tilde{j}_F : T_F^* M \to T^*F$.

Let us set $h_F = \tilde{j}_F \circ i_F : T_F M \to T^*F$. The kernel of h_F is $T_F N$ and we have

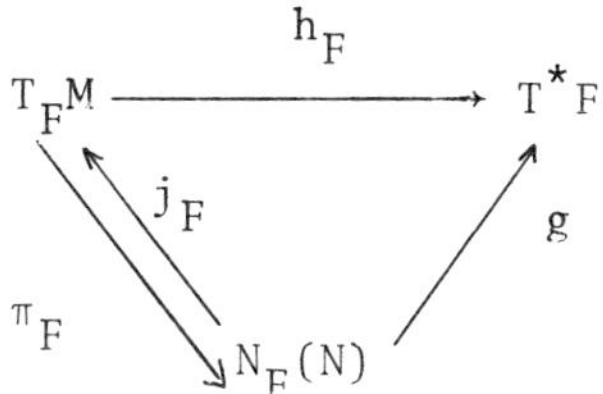

where $N_F(N) = \frac{T_N M}{TN} |F$, π_F is induced by the projection $T_N M \to \frac{T_N M}{TN}$, j_F is a relevement of π_F and g is an isomorphism.

Now we set

$$Q_F = Q^F \circ j_F \circ g^{-1}.$$

Obviously all the above discussion is invariant under the action of a canonical isomorphism.

Finally we set

$$q_F = Q_F + p_s|_F \tag{2.3}$$

where p_s is the subprincipal symbol of P which is invariantly defined on F.

Now if $f \in C^\infty(\Omega)$ we set $\tilde{f} = f \circ p_o \circ \mathfrak{I}_F \circ \tilde{\pi}_F$ where $\tilde{\pi}_F : T^*F \to F$, $\mathfrak{I}_F : F \to T^*\Omega$, $p_o : T^*\Omega \to \Omega$.

We can now state the

<u>Definition 2.1</u> :

<u>Let</u> $(x_0, y_0) \in \Omega$, $f \in C^\infty(\Omega)$ <u>such</u> <u>that</u> $df(x_0,y_0) \neq 0$.

The non-characteristic hypersurface $\{f(x,y)=f(x_o,y_o)\}$ is called pseudo-convex, with respect to the bicharacteristics of P, at (x_o,y_o) if

$$(\psi)\begin{cases} \text{For every } z_o\in N \text{ such that } \pi(z_o)=(x_o,y_o) \text{ there exists } F\subset N \text{ such that :} \\ (z_o,\zeta_o)\in T^*F,\ q_F(z_o,\zeta_o)=0 \text{ and } H_{q_F}(\tilde{f})(z_o,\zeta_o)=0 \\ \Longrightarrow H^2_{q_F}(\tilde{f})(z_o,\zeta_o)>0 \end{cases}$$

where H_{q_F} is the Hamiltonian of q_F defined in (2.3).

The same hypersurface will be called strongly non-pseudo-convex with respect to the bicharacteristics of P if

$$(\not\psi)\begin{cases} \exists\ z_o\in N:\ \pi(z_o)=(x_o,y_o),\ \exists\ F\ni z_o, \exists\,\zeta_o\in T^*_{z_o}F \text{ such that} \\ q_F(z_o,\zeta_o)=0,\ H_{q_F}(\tilde{f})(z_o,\zeta_o)=0 \text{ and } H^2_{q_F}(\tilde{f})(z_o,\zeta_o)<0. \end{cases}$$

Example 2.2 :

In the coordinates $(x,t,y)\in\mathbb{R}^{n-1}\times\mathbb{R}\times\mathbb{R}$ let us suppose that

(2.4) $$P = D_t^2 + \sum_{|\alpha|\leq 2} a_\alpha(x,t,y)D_x^\alpha + c(x,t;y)D_y,$$

$(x_o,t_o,y_o) = (0,0,0)$ and

$$S = \{(x,t,y) : t = 0\}.$$

Here $N = \{(x,t,y,\xi,\tau,\eta) : \tau = \xi = 0\}$. On the other hand, we have

$F = \{(x,t,y,\xi,\tau,\eta): y=y_o,\ \xi=0,\ \tau=0,\ \eta=\eta_o,\ x\in\mathbb{R}^{n-1},\ t\in\mathbb{R}\},\ \eta_o\neq 0.$

A point (z,ζ) in T^*F can be written $(x,t,y_o,0,0,\eta_o;\ \zeta_1,\zeta_2,0,0,0,0)$.

So

$$q_F(z,\zeta) = \zeta_2^2 + a(x,t,y_o;\zeta_1) + c(x,t,y_o)\eta_o.$$

Now since $\tilde{f} \equiv t$ we have $\{q_F, \tilde{f}\} = -\frac{\partial}{\partial \zeta_2} q_F = -2\zeta_2$ and $\{q_F, \{q_F, \tilde{f}\}\} = 2\frac{\partial q_F}{\partial t}$. The condition (ψ) says:

for every $\zeta_1 \in \mathbb{R}^{n-1} \setminus 0$ such that $a(0,0,0;\zeta_1) + c(0,0,0)\eta_o = 0$ we have $\frac{\partial a}{\partial t}(0,0,0;\zeta_1) + \frac{\partial c}{\partial t}(0,0,0).\eta_o > 0$. This is equivalent to

$$(2.5) \quad \forall\ \xi \in \mathbb{R}^{n-1} \setminus 0, \ \frac{\partial a}{\partial t}(0;\xi) - \frac{\frac{\partial c}{\partial t}(0)}{c(0)}\, a(0;\xi) > 0 .$$

The condition $(\not\psi)$ reads

$$(2.6) \quad \exists\ \xi^o \in \mathbb{R}^{n-1} \setminus 0 : \ \frac{\partial a}{\partial t}(0,\xi^o) - \frac{\frac{\partial c}{\partial t}(0)}{c(0)}\, a(0;\xi^o) < 0 .$$

Let us now state the uniqueness result.

<u>Theorem 2.3</u> :

<u>Let</u> Ω <u>be an open set in</u> $\mathbb{R}^{n+1}$, $(x_o,y_o) \in \Omega$ <u>and</u> P <u>be the differential operator</u>

$$P = \sum_{|\alpha| \leq 2} a_\alpha(x,y) D_x^\alpha + c(x,y) D_y .$$

<u>We suppose that the principal symbol of</u> P <u>and</u> c <u>are real and</u> C^∞ <u>near</u> (x_o,y_o) <u>and that the lower order terms are bounded</u>.
<u>Let</u> $\{f(x,y) = f(x_o,y_o)\}$ <u>be a hypersurface which is pseudo-convex with respect to the bicharacteristics of</u> P <u>at</u> (x_o,y_o). <u>Then there exists a neighborhood</u> ω <u>of</u> (x_o,y_o) <u>such that if</u> $u \in C^\infty(\Omega)$ <u>satisfies</u> $Pu = 0$ <u>in</u> ω <u>and has its support contained in</u> $\{(x,y) \in \omega : f(x,y) \geq f(x_o,y_o)\}$, <u>then</u> u <u>vanishes in</u> ω .

<u>Examples 2.4</u> : The following examples are written in the (x,t,y) - coordinates, $x \in \mathbb{R}^{n-1}$, $t \in \mathbb{R}$, $y \in \mathbb{R}$ and $f(x,t,y) = t$. They are particular cases of the above result.

$$(1) \qquad P = D_t^2 + \varepsilon_1 \sum_{j=1}^{n-1} D_{x_j}^2 + (1+\varepsilon_2 t)D_y \ , \ \varepsilon_1\varepsilon_2 = 1$$

$$(2) \qquad P = D_t^2 + t \sum_{j=1}^{n-1} D_{x_j}^2 + D_y$$

$$(3) \qquad P = D_t^2 + (1+\alpha t) \sum_{j=1}^{n-1} D_{x_j}^2 + (1+\beta t)D_y \ , \ \alpha > \beta .$$

Proof of Theorem 2.3 :

Using a suitable change of coordinates we can assume without loss of generality that

$$P = D_t^2 + a(x,t,y;D_x) + c(x,t,y)D_y + \sum_{j=1}^{n-1} \alpha_j D_{x_j} + \alpha D_t + \gamma$$

where $a(x,t,y;\xi)$ is real quadratic form, c is real, $c(0,0,0) \neq 0$, $f(x,y,t) = t$ and

$$(2.7) \qquad \frac{\partial a}{\partial t}(0;\xi) - \frac{\frac{\partial c}{\partial t}(0)}{c(0)} a(0;\xi) > 0 \quad \forall \ \xi \in R^{n-1} \setminus 0 .$$

By the classical Holmgren transformation $(x,t,y)\,(x, t+\delta(|x|^2+y^2), y)$ with a small $\delta > 0$ we make the support of the solution contained in the convex set $\{(x',t',y') : t' > \delta(|x'|^2 + y'^2)\}$. The operator P becomes

$$(2.8) \quad Q = D_{t'}^2 + \tilde{a}(x',t',y'; \tfrac{1}{i} X) + \tilde{c}(x',t',y')D_{y'} +$$

$$+ \sum_{j=1}^{n-1} \tilde{\alpha}_j D_{x_j'} + \tilde{\alpha} D_{t'} + \tilde{\beta}$$

where $X = (X_1,\ldots,X_n)$, $X_j = \frac{\partial}{\partial x_j'} + 2\delta x_j' \frac{\partial}{\partial t}$ and $\tilde{a}$, $\tilde{c}$ satisfy also (2.7).

In the following we shall use (x,y,t) instead of (x',y',t'). Now theorem 2.3 will follow as usual from the Carleman estimate which we now state.

Proposition 2.5 :

Let Q be defined by (2.8) with $\tilde{a}$ and $\tilde{c}$ satisfying (2.7). There exist positive constants C, γ_o, T_o, r such that, for $u \in C^\infty(V)$ with $\operatorname{supp} u \subset \{0 \leq t \leq T_o, |x| + |y| \leq r\}$ and for $\gamma \geq \gamma_o$, we have

$$(2.9) \quad \left\{ \begin{array}{l} \gamma^3 \| t^{-\gamma+\frac{3}{2}} u \|^2_{L^2} + \gamma \sum_{j=1}^{n-1} \| t^{-\gamma} D_{x_j} u \|^2_{L^2} + \\ + \gamma \| t^{-\gamma} D_t u \|^2_{L^2} \leq C \| t^{-\gamma+\frac{1}{2}} Qu \|^2_{L^2} . \end{array} \right.$$

Proof :

It is of course equivalent to prove (2.9) for the operator

$$Q_o = \frac{\partial^2}{\partial t^2} + \sum_{i,j=1}^{n-1} X_j \, a_{ij} \, X_i - c(x,t,y) D_y .$$

Setting $v = t^{-\gamma} u$ we see easily that $t^{-\gamma} Q_o u = \tilde{Q} v$ with

$$Q = \partial_t^2 + \sum_{i,j=1}^{n-1} X_j \, a_{ij} \, X_i + \gamma(\gamma-1) t^{-2} d + 2\gamma t^{-1} \partial_t + 2\delta\gamma t^{-1} Z - cD_y$$

where $Z = (\sum_{i,j=1}^{n-1} a_{ij} x_i X_j + \sum_{i,j=1}^{n-1} X_j a_{ij} x_i)$ so $Z^* = -Z$, $d \in C^\infty$,

$d(0) = 1$.

Let λ be a real number such that $\lambda > \left| \frac{\frac{\partial c}{\partial t}(0)}{c(0)} \right|$ and let $Q_1 = \tilde{Q} + \lambda \frac{\partial}{\partial t}$; then (2.9) will be a consequence of the following inequality

$$(2.10) \quad \gamma^3 \| t^{-\frac{3}{2}} v \|^2 + \gamma \sum_{j=1}^{n-1} \| X_j v \|^2 + \gamma \| D_t v \|^2 \leq C \| t^{\frac{1}{2}} Q_1 v \|^2 .$$

Let us set $t^{\frac{1}{2}} Q_1 = t^{\frac{1}{2}} A + t^{\frac{1}{2}} B$ with

$$(2.11)\quad \begin{cases} t^{\frac{1}{2}}A = t^{\frac{1}{2}}\partial_t^2 + t^{\frac{1}{2}}\sum_{i,j} X_j a_{ij} X_i + \gamma(\gamma-1)t^{-\frac{3}{2}} - ct^{\frac{1}{2}}D_y + \lambda t^{\frac{1}{2}}\partial_t \\ t^{\frac{1}{2}}B = 2\gamma t^{-\frac{1}{2}}\partial_t + 2\delta\gamma t^{-\frac{1}{2}}Z . \end{cases}$$

Then

$$(2.12)\quad ||t^{\frac{1}{2}}Q_1 v||^2 = ||t^{\frac{1}{2}}Av||^2 + ||t^{\frac{1}{2}}B_v||^2 + \mathrm{Re}(t^{\frac{1}{2}}Av, t^{\frac{1}{2}}Bv) .$$

Now, we see easily that

$$(2.13)\quad 2\mathrm{Re}(t^{\frac{1}{2}}\partial_t^2 v,\ 2\gamma t^{-\frac{1}{2}}\partial_t v) = 0$$

$$(2.14)\quad 2\mathrm{Re}(t^{\frac{1}{2}}\sum_{i,j} X_j a_{ij} X_i v, 2\gamma t^{-\frac{1}{2}}\partial_t v) = 2\gamma \sum_{i,j}(\frac{\partial a_{ij}}{\partial t} X_i v, X_j v)$$

$$(2.15)\quad \begin{cases} 2\mathrm{Re}(\gamma(\gamma-1)t^{-3/2}dv,\ 2\gamma t^{-\frac{1}{2}}\partial_t v) = 4\gamma^2(\gamma-1)||t^{-3/2}\sqrt{d}\,v||^2 + R_1 \\ |R_1| \leqslant C\ \gamma^2(\gamma-1)||t^{-1}v||^2. \end{cases}$$

Now we have

$$(2.16)\quad \begin{cases} 2\mathrm{Re}(-ct^{\frac{1}{2}}D_y v,\ 2\gamma t^{-\frac{1}{2}}\partial_t v) = 2\gamma(cD_y v, \frac{c'_t}{c}v) + R_2 \\ |R_2| \leqslant C_1(||\partial_t v||^2 + \gamma^2||v||^2). \end{cases}$$

It follows from (2.11) that

$$(2.17)\quad cD_y v = \partial_t^2 + \sum X_j a_{ij} X_i + \gamma(\gamma-1)t^{-2} + \lambda\partial_t - A .$$

Using (2.16) and (2.17) we see easily that

$$(2.18)\quad \begin{aligned} &2\mathrm{Re}(-ct^{\frac{1}{2}} D_y v,\ 2\gamma t^{-\frac{1}{2}}\partial_t v) = \\ &= -2\gamma \sum_{i,j}(\frac{c'_t}{c} a_{ij} X_i v, X_j v) - 2\gamma(\partial_t v, \frac{c'_t}{c}\partial_t v) + R_3 \end{aligned}$$

with

$$\forall\ \varepsilon>0\ \exists C_\varepsilon>0\ :\ |R_3|\leqslant \varepsilon\,||t^{\frac{1}{2}}Av||^2+C_\varepsilon\gamma^2||t^{-\frac{1}{2}}v||^2+ \tag{2.19}$$
$$+\ C_0(||\partial_t v||^2+\sum_j ||X_j v||^2+\gamma^3||t^{-1}v||^2).$$

Now

$$2\mathrm{Re}(\lambda t^{\frac{1}{2}}\partial_t v,\ 2\gamma t^{-\frac{1}{2}}\partial_t v) = 2\gamma\lambda||\,\partial_t v||^2 \tag{2.20}$$

$$|2\mathrm{Re}(t^{\frac{1}{2}}\partial_t v,\ 2\delta\gamma t^{-\frac{1}{2}}Zv)|\leqslant C\delta\gamma(||\partial_t v||^2+\sum_j ||X_j v||^2+||v||^2) \tag{2.21}$$

$$2\delta\gamma|2\mathrm{Re}(\sum_{i,j} X_j a_{ij}X_i v,\ Zv)|\leqslant C\delta\gamma(\sum_j ||X_j v||^2+\ ||v||^2) \tag{2.22}$$

$$2\delta\gamma^2(\gamma-1)|\ 2\mathrm{Re}\,(dt^{-\frac{3}{2}}v,\ t^{-\frac{1}{2}}Zv)|\leqslant C\ \delta\gamma^3\ ||t^{-\frac{3}{2}}v||^2 \tag{2.23}$$

$$\begin{cases} 2\delta\gamma|2\mathrm{Re}(-cD_y v,\ Zv)|\leqslant \varepsilon\ ||t^{\frac{1}{2}}Av||^2+C_\varepsilon\ \gamma^2\ ||t^{-\frac{1}{2}}v||^2\ + \\ +\ C\delta(\gamma||\,\partial_t v||^2+\gamma\sum_j ||X_j v||^2+\gamma^3||\,t^{-3/2}v||^2). \end{cases} \tag{2.24}$$

Finally

$$2\delta\gamma\lambda\,|\,2\mathrm{Re}(\partial_t v, Zv)\,|\leqslant C\delta\gamma\ (||\partial_t v||^2+\sum_j ||X_j v||^2+||v||^2). \tag{2.25}$$

Using (2.7) and (2.12) to (2.25), taking first ε small enough, then choosing T_0, r, δ and γ_0^{-1} small enough and $\lambda>|\frac{c'_t}{c}(0)|$, we deduce (2.10).

We are going now to state the first non-uniqueness result.

<u>Theorem 2.6</u> :

<u>With the notations of theorem</u> 2.4, <u>let</u> $S=\{f(x,y) = f(x_0,y_0)\}$ <u>be a hypersurface strongly non-pseudo-convex, with respect to the bicharacteristics of</u> P, <u>at</u> (x_0,y_0).

<u>Then in every neighborhood</u> ω <u>of</u> (x_o,y_o). <u>we can find</u> $(x_1,y_1) \in S \cap \omega$, <u>a neighborhood</u> ω_1 <u>of</u> (x_1,y_1), <u>two functions</u> u <u>and</u> b <u>in</u> $C^\infty(\omega_1)$ <u>with support in</u> $\{(x,y) \in \omega : f(x,y) \geq f(x_1,y_1)\}$ <u>such that</u> :

$$\begin{cases} Pu + bu = 0 \\ (x_1,y_1) \in \operatorname{supp} u . \end{cases}$$

<u>Remark 2.7</u> :

Let us consider in the (x,t,y) coordinates the operator

$$P = D_t^2 + a(x,t,y;D_x) + c(x,t,y)D_y$$

and $(x_o,t_o,y_o) = (0,0,0)$, $f(x,y,t) = t$. According to (2.6) the strongly non-pseudo-convexity condition is

$$\exists\ \xi^o \in \mathbb{R}^{n-1}\setminus 0 : \frac{\partial a}{\partial t}(0,\xi^o) - \frac{\frac{\partial c}{\partial t}(0)}{c(0)}\ a(0;\xi^o) < 0 .$$

If $a(0;\xi^o) \neq 0$ or if $a(x,t,y;\xi) = t^\ell\ \overline{a}(x,t,y;\xi)$, $\ell \in \mathbb{N}^*$, then in the above theorem we can take $(x_1,t_1,y_1) = (0,0,0)$ and $\omega_1 = \omega$.

<u>Examples 2.8</u> :

The following examples which, according to Remark 2.7 are particular cases of theorem 2.6, must be compared with Examples 2.4

(1) $$P = D_t^2 + \varepsilon_1 \sum_{j=1}^{n-1} D_{x_j}^2 + (1+\varepsilon_2 t)D_y,\ \varepsilon_1\varepsilon_2 = -1$$

(2) $$P = D_t^2 - t \sum_{j=1}^{n-1} D_{x_j}^2 + D_y$$

(3) $$P = D_t^2 + (1+\alpha t) \sum_{j=1}^{n-1} D_{x_j}^2 + (1+\beta t)D_y,\ \alpha < \beta .$$

Proof of Theorem 2.6 :

We shall work in the coordinates where

$$P = D_t^2 + a(x,t,y;D_x) + c(x,t,y)D_y + \sum_{j=1}^{n-1} \alpha_j D_{x_j} + \alpha D_t + \beta$$

and $f(x,t,y) = t$. Let us recall that $c(0) \neq 0$ and

$$(2.26) \quad \exists\ \xi^o \in \mathbb{R}^{n-1} : \frac{\partial a}{\partial t}(0;\xi^o) - \frac{\frac{\partial c}{\partial t}(0)}{c(0)} a(0,\xi^o) < 0 .$$

Lemma 2.9 :

There exists $(x_1,y_1) \in V \cap \{t=0\}$, $\tilde{\xi}_o \in \mathbb{R}^{n-1} \setminus 0$, $\bar{a}(x,t,y;\xi)$ such that

$$(2.27) \quad \begin{aligned} & a(x,t,y;\xi) = t^{\ell}\, \bar{a}(x,t,y;\xi), \quad \forall (x,t,y;\xi) \in V \times \mathbb{R}^{n-1}, \ \ell = 0 \text{ or } 1 \\ & \bar{a}(x_1,0,y_1;\tilde{\xi}_o) \neq 0 \end{aligned}$$

$$(2.28) \quad \frac{\partial a}{\partial t}(x_1,0,y_1;\tilde{\xi}_o) - \frac{\frac{\partial c}{\partial t}}{c}(x_1,0,y_1)\ a(x_1,0,y_1;\tilde{\xi}_o) < 0 .$$

Proof :

If $a(0;\xi^o) \neq 0$ we take $(x_1,y_1) = (0,0)$, $\ell = 0$ and $\bar{a} = a$. Let us suppose that $a(0,\xi^o) = 0$. Without loss of generality we can assume that $\xi^o = (1,0,\ldots,0)$.

Let $W \times \Gamma_{\xi^o}$ be a conic neighborhood of $(0,0,\xi^o)$ in $\mathbb{R}_x^{n-1} \times \mathbb{R}_y^n \times \mathbb{R}_\xi^{n-1}$ such that

$$\frac{\partial a}{\partial t}(x,0,y;\xi) - \frac{\frac{\partial c}{\partial t}}{c}(x,0,y)\ a(x,0,y;\xi) < 0$$

for every $(x,y,\xi) \in W \times \Gamma_{\xi^o}$.

Setting $a(x,t,y;\xi) = \sum_{i,j=1}^{n-1} a_{ij}(x,t,y)\xi_i\xi_j$ let us suppose $a_{11}(x,0,y) \neq 0$ in W ; we can find (x_1,y_1) in W such that $a_{11}(x_1,0,y_1) \neq 0$; then $a(x_1,0,y_1;\xi_o) = a_{11}(x_1,0,y_1) \neq 0$, i.e. $\tilde{\xi}_o = \xi^o$, $\bar{a} = a$ and $\ell = 0$. If $a_{11}(x,0,y) \equiv 0$ in W

then $a_{11}(x,t,y) = t^{\ell_1}\tilde{a}_{11}$ with $\ell_1 \geq 1$. If there exist $(x_1,y_1)\in W$ and $2 \leq i \leq n-1$ such that $a_{1i}(x_1,0,y_1) \neq 0$, we take $\tilde{\xi}_o = (1,0,0,\varepsilon,0,\ldots,0) \in \Gamma_{\xi^o}$, $\bar{a} = a$, $\ell = 0$; if not, we get $a_{1i}(x,t,y) = t^{\ell_{1i}}\tilde{a}_{1i}, \ell_{1i} \neq 0$. If there exist $2 \leq i \leq n-1$ and $(x_1,y_1) \in W$ such that $a_{ii}(x_1,0,y_1) \neq 0$, then we take $\tilde{\xi}_o = (1,0,.,0,\varepsilon,.,0)$, $a=a$, $\ell=0$; if not, then $a_{ii}(x,t,y) = t^{\ell_i}\tilde{a}_{ii}$, $\ell_i \neq 0$. Finally, if there exist (i,j), $i \neq j$, and $(x_1,y_1)\in W$ such that $a_{ij}(x_1,0,y_1)\neq 0$, then we set $\tilde{\xi}_o = (\tilde{\xi}_k)$, $\tilde{\xi}_1=1$, $\tilde{\xi}_i=\tilde{\xi}_j=\varepsilon$, $\tilde{\xi}_k=0$ otherwise ; if not, we have $a_{ij} = t^{\ell_{ij}}\tilde{a}_{ij}$, $\ell_{ij} \neq 0$. This implies that $a(x,t,y;\xi) = t^{\ell}\bar{a}(x,t,y;\xi)$, $\ell \geq 1$; it follows from (2.26) that $\ell=1$ and $\bar{a}(0;\xi^o) < 0$. ■

In the following we shall suppose that $(x_1,y_1)=(0,0)$ in Lemma 2.9. The construction of b and u is that used in the proof of theorem 3.1 Chapter One.

So we omit here the details and refer the reader to that proof for them.

Let $\delta > 0$, $\theta \in N^*$, $\lambda = \delta^{-\theta}$ and $t-\delta = \frac{s}{\lambda}$. We have

$$P=\lambda^2 D_s^2 + a(x,\delta+\frac{s}{\lambda},y;D_x)+c(x,\delta+\frac{s}{\lambda},y)D_y + \sum_{j=1}^{n-1}\tilde{\alpha}_j D_{x_j} + \tilde{\alpha}\lambda D_s + \tilde{\beta}$$

where $\tilde{\alpha}_j = \alpha_j(x,\delta+\frac{s}{\lambda},y)$ etc...

The formal solutions of $Pu = 0$ will be

$$u(x,s,y,\delta)= e^{i(\sqrt{\tau}\,\xi(x,y,\delta)+\tau\psi(\delta)y)}e^{\nu\phi(x,s,y,\delta)-\gamma(x,\delta,y)}w(x,s,y,\delta)$$

where $\tau = \lambda^2\nu^2$ and ξ,ϕ have to satisfy the "phase equation"

$$(2.30)\quad -\left(\frac{\partial\phi}{\partial s}\right)^2 + a(x,\delta+\frac{s}{\lambda},y;\nabla_x\xi(x,y,\delta)) + \psi(\delta)\,c(x,\delta+\frac{s}{\lambda},y) = 0.$$

Proposition 2.10 :

We can find a neighborhood ω of $(x,y) = (0,0)$, positive constants δ_o, s_o, three real functions $\xi(x,y,\delta) \in C^\infty(\omega \times]0,\delta_o])$, $\phi \in C^\infty(\omega \times]-s_o,s_o[\times [0,\delta_o])$ and $\psi(\delta) \in C^\infty(]0,\delta_o])$ satisfying the phase equation (2.30), such that

$$(2.31)\quad \begin{cases} |\xi(x,y,\delta)| \leqslant C\delta^{-M_1} \text{ for } (x,y,\delta) \in \omega \times]0,\delta_o] \\ |\psi(\delta)| \leqslant C\delta^{-M_2}, \ \delta \in]0,\delta_o] \\ \phi(x,y,s,\delta) = s - \beta(x,y,s,\delta)s^2, \ \beta(0) > 0. \end{cases}$$

Proof :

Let us denote by B^∞ the set of functions $f = f(x,y,s,\delta)$ which are C^∞ in (x,y,s) and satisfy $|D_x^\alpha D_y^\beta D_s^\ell f| \leqslant C_{\alpha\beta\ell}$ for $(x,y,s,\delta) \in \omega \times]-s_o,s_o[\times]0,\delta_o]$.

We first show that we can find $B = B(x,y,\delta) \subset B^\infty$, $\psi(\delta)$ and $\xi(x,y,\delta)$ in $C^\infty(\omega \times]0,\delta_o])$ such that

$$(2.32)\quad \begin{cases} a(x,\delta,y;\nabla_x\xi) + \psi(\delta)\, c(x,\delta,y) = 1 \\ \frac{\partial a}{\partial t}(x,\delta,y;\nabla_x\xi) + \psi(\delta)\frac{\partial c}{\partial t}(x,\delta,y) = -\lambda B(x,y,\delta) \end{cases}$$

with $B(0,0,0) > 0$, $\xi = \delta^{-\frac{1}{2}\theta}\tilde{\xi}$, $\tilde{\xi} \in B^\infty$ and $\psi(\delta) = \delta^{\ell-\theta}\psi_o, \psi_o \in \mathbb{R}$ $\ell = 0,1$ being defined in Lemma 2.9.

Using (2.27) and setting $\xi = \delta^{-\frac{\theta}{2}}\cdot\tilde{\xi}$, the first equation of (2.32) becomes

$$\bar{a}(x,\delta,y\ ;\nabla_x\tilde{\xi}) = -\delta^{\theta-\ell}\psi(\delta)\, c(x,\delta,y) + \delta^{\theta-\ell}.$$

Let us set $\psi(\delta) = \delta^{\ell-\theta}\psi_o, \psi_o$ to be chosen. Then

$$\bar{a}(x,\delta,y;\nabla_x\tilde{\xi}) = -\psi_o\, c(x,\delta,y) + \delta^{\theta-\ell}.$$

We chose $\psi_o = - \frac{\bar{a}(0;\tilde{\xi}_o)}{c(0)}$ where $\tilde{\xi}_o$ is defined in lemma 2.9.

By Euler's relation we can find j such that $\frac{\partial \bar{a}}{\partial \xi_j}(0;\tilde{\xi}_o) \neq 0$.

So we can solve, by the Hamilton-Jacobi theory, the problem

$$\begin{cases} \bar{a}(x,\delta,y;\nabla_x \tilde{\xi}) = -\psi_o\, c(x,\delta,y) + \delta^{\theta-\ell} \\ \tilde{\xi}|_{x_j=0} = \sum_{i\neq j} x_i\, \tilde{\xi}_{o,i} \,. \end{cases}$$

We find a solution $\tilde{\xi} \in C^\infty$ such that $\nabla_x \tilde{\xi}(0) = \xi_o$.

Let us set

$$B = -\left(\frac{\partial a}{\partial t}(x,\delta,y;\nabla_x \tilde{\xi}) + \delta^\theta \psi(\delta)\, \frac{\partial c}{\partial t}(x,\delta,y)\right)$$

then

$$B = -\left(\frac{\partial a}{\partial t}(x,\delta,y;\nabla_x \tilde{\xi}) - \frac{\delta^\ell \bar{a}(0,\tilde{\xi}_o)}{c(0)}\, \frac{\partial c}{\partial t}(x,\delta,y)\right)$$

$$B = -\left(\frac{\partial a}{\partial t}(x,\delta,y;\nabla_x \tilde{\xi}) - \frac{\delta^\ell \bar{a}(x,\delta,y;\tilde{\xi}_o)\, \frac{\partial c}{\partial t}(x,\delta,y)}{c(0)} + b(x,y,\delta,\tilde{\xi}_o)\right)$$

where $b(0,0,0;\tilde{\xi}_o) = 0$. Since $\delta^\ell \bar{a} = a$, we get by (2.28)

$$B(0) = -\left(\frac{\partial a}{\partial t}(0;\tilde{\xi}_o) - \frac{\frac{\partial c}{\partial t}}{c}(0)\; a(0;\tilde{\xi}_o)\right) > 0 .$$

Now using (2.32) we can write the phase equation (2.30)

$$\left(\frac{\partial \phi}{\partial s}\right)^2 = 1 - sB(x,y,\delta) + \frac{s^2}{\lambda^2} \int_0^1 (1-u)\left[a''_{tt}(x,\delta+u\tfrac{s}{\lambda},y;\nabla_x \xi) + \psi(\delta)\, c''_{tt}(x,\delta+u\tfrac{s}{\lambda},y)\right] du .$$

The coefficient of s^2 in the above equation is in B^∞ since the singularity in δ coming from ξ and ϕ is destroyed by $\frac{1}{\lambda^2} = \delta^{2\theta}$.
It follows that there exists a solution ϕ of the form

$$\phi = s - \frac{1}{4} B(x,y,\delta) s^2 + o(s^2) .$$

Let us now give the expression of the parameters.

Let ε be in $]0,\frac{1}{10}[$. We set $\theta = 1+\varepsilon$, $\rho = \frac{1-\varepsilon}{\varepsilon}$, $\delta = b_k = k^{-\rho}$, $\lambda_k = \delta^{-\theta} = k^{\rho\theta}$, $\nu_k = k^{\sigma}$ with $\sigma = (\frac{1}{2} - \frac{3}{2}\varepsilon)$, $\tau_k = \lambda_k^2 \nu_k^2$.

Then $\gamma_k(x,y) \sim \frac{\rho}{\sigma-\varepsilon+1} k^{\sigma-\varepsilon+1}$, and setting

$$F_k(x,y,t) = \mathrm{Log}\left|\frac{v_k(x,y,t)}{v_{k+1}(x,y,t)}\right|, \text{ we get } \frac{\partial F_k}{\partial t} \geqslant Ck^{\eta} \text{ with } \eta > \rho+1,$$

which makes the construction of u and b possible as in theorem 3.1 Chapter One.

To end this section we discuss an example where the initial hypersurface is neither pseudo-convex nor strongly non-pseudo-convex.

Theorem 2.10 :

Let us consider in a neighborhood of the origin in $\mathbb{R}^{n+1}$ *the operator with* C^{∞} *coefficients*

$$P = D_t^2 + \sum_{|\alpha| \leqslant 2} a_{\alpha}(x,y) D_x^{\alpha} + c(x,y) D_y$$

where $a_{\alpha}, |\alpha|=2$, *and* c *are real and* $\sum_{|\alpha|=2} |a_{\alpha}(0,0)| \neq 0$.

There exist two functions b *and* u, C^{∞} *near the origin, with support in* $\{(x,y,t); t \geqslant 0\}$, *such that*

$$\begin{cases} Pu + bu = 0 \\ (0,0,0) \in \operatorname{supp} u . \end{cases}$$

Sketch of proof :

Let us set $a^{o}(x,y;\xi) = \sum_{|\alpha|=2} a_{\alpha}(x,y)\xi^{\alpha}$; then there exists $\xi^o \in \mathbb{R}^{n-1}\setminus 0$ such that $a^o(0,0;\xi^o) \neq 0$.

Lemma 2.11 :

Let us set $\varepsilon = -\operatorname{sign}\left(\frac{c(0,0)}{a(0,0;\xi^o)}\right)$. *We can find a* C^∞ *function* $\xi(x,y)$ *near the origin such that*

$$(2.33)\qquad \begin{cases} a^o(x,y;\nabla_x\xi(x,y)) + \varepsilon\, c(x,y)=0 \\ \nabla_x\xi(0,0) \neq 0. \end{cases}$$

Proof :

By Euler's relation we can find $1 \leqslant j \leqslant n-1$ such that $\frac{\partial a^o}{\partial \xi_j}(0,0;\xi^o) \neq 0$, so using Hamilton-Jacobi theory we can solve the problem

$$\begin{cases} a^o(x,y;\nabla_x\xi(x,y)) + \varepsilon\, c(x,y) = 0 \\ \xi\big|_{x_j=0} = \sum\limits_{i\neq j} x_i\, \tilde{\xi}_i^o \end{cases}$$

where $\tilde{\xi}_i^o = \sqrt{-\varepsilon\frac{c(0,0)}{a(0,0;\xi^o)}}\,\xi_i^o$, since then $\nabla_x\xi(0,0) = \tilde{\xi}_o$ and

$a^o(0,0;\tilde{\xi}_o) + \varepsilon\, c(0,0) = 0$.

Now setting $t-\delta = \frac{s}{\lambda}$, $\lambda=\delta^{-\theta}$, $\delta=b_k$ we set for $t \in [b_{k+1},b_k]$

$$u_k(x,y,s) = e^{i(\sqrt{\tau}_k\,\xi(x,y) + \varepsilon\tau_k y)}\, e^{-\nu_k\phi_k(s)}\, e^{-\gamma_k}\, w_k(x,y,s).$$

where ξ and ε have been defined above. We take $\theta=2$, $b_k = k^{-1}$, $\phi_k(s) = \psi(s).s$ where ψ is a real and C^∞ function such that

$\psi = -\frac{1}{2}$ for $s \leqslant -\frac{1}{8}$, $\psi = 1$ for $s \geqslant \frac{1}{8}$, $\nu_k = k^{-2}$, $\gamma_k = k^{-2}$, $\tau_k = k^{-2\rho}$ with ρ big enough. The function w is obtained by transport equations which are here tangential. The scheme of the proof is similar to that of theorem 3.1 Chapter One. The details are left to the reader.

§3. THE CASE OF DOUBLE REAL ROOTS WITH CONSTANT MULTIPLICITY

As we have seen before, in the classical results of uniqueness, the complex roots (in the sense of the Cauchy problem) are allowed to be muliple but the real ones are assumed to be simple.
We shall consider here the case where there is one double real root of constant multiplicity.

Let us consider in $\mathbb{R}^{n+1} = \{(x,t) : x \in \mathbb{R}^{n-1}, t \in \mathbb{R}\}$ the operator

$$P = -D_t^2 + p_1(x,t;D_x) + ip_2(x,t;D_x) + \alpha D_t + \beta \tag{3.1}$$

where $p_j(x,t,\xi) = \sum_{k=1}^{n} a_{jk}(x,t)\xi_k$ has real coefficients which are C^∞ in a neighborhood V of the origin.
The initial hypersurface will be $\{(x,t) : t=0\}$. We shall denote by V_+ the set $\{(x,t) \in V : t=0\}$.

<u>Theorem 3.1:</u>

<u>Let</u> $P_o = \frac{\partial^2}{\partial t^2} + p_1 + ip_2$. <u>Let us suppose that there exist bounded functions in</u> V^+, a,b,c,f,g <u>such that</u> $t\frac{\partial b}{\partial t}$ <u>is bounded and</u>

$$\left[\frac{\partial}{\partial t}, p_2\right] = \frac{a}{t} p_2 \tag{H.1}$$

(H.2) $$\left[\frac{\partial}{\partial t}, p_1\right] = \frac{b}{t} p_1 + \frac{c}{t} p_2$$

(H.3) <u>there exists</u> $\varepsilon > 0$ <u>such that in</u> V_+

$$\begin{cases} b(x,t) + 2 \geqslant \varepsilon \\ 2(a(x,t)+1) - b(x,t) \geqslant \varepsilon \\ (b(x,t)+2)(2a(x,t) + 2-b(x,t)) \geqslant c^2(x,t) \end{cases}$$

(H.4) $$[p_1, p_2] = fp_1 + gp_2 .$$

<u>Then there exists a neighborhood</u> W <u>of the origin in which every</u> $u \in C^\infty(V)$, <u>satisfying with</u> $C=C(u)$

(3.2) $$\begin{cases} |P_o u(x,t)| \leqslant C\{\frac{1}{t}\left|\frac{\partial u}{\partial t}(x,t)\right| + \frac{1}{t^2}|u(x,t)|\}, (x,t)\in V_+ \\ u_{|t<0}=0 , \end{cases}$$

<u>must vanish.</u>

<u>Examples 3.2:</u>

The following operators satisfy conditions (H.i) i = 1,...,4, of theorem 3.1.

a) In $\mathbb{R}^2$, $P_o = D_t^2 + (\alpha(x,t) + i\beta(x,t))t^\ell D_x$ where α and β are real and C^∞ such that $\alpha(0,0)\neq 0$ (or $\alpha \equiv 0$), $\beta(0,0)\neq 0$ (or $\beta \equiv 0$) and $\ell \in \mathbb{N}$.

b) In $\mathbb{R}^2$, $P_o = D_t^2 + \alpha(x,t)(t+x^2)D_x$, α real C^∞, $\alpha(0,0) \neq 0$.

c) In $\mathbb{R}^3$, $P_o = D_t^2 + t^b\alpha D_1 + it^a\beta D_2$, where α and β are non-vanishing real smooth functions and a,b integers such that $2(a+1) > b$.

d) In $\mathbb{R}^3, P_o = D_t^2 + t^\ell(D_1 + dt^m D_2) + it^{\ell+m} D_2$ where ℓ , m belong to $\mathbb{N}$, d is real number and $(\ell + 2)(2m+\ell+2) \quad m^2 d^2$.

The conditions involved in theorem 3.1 seem to be quite restrictive, but as we shall see in the second part of this section all of them (except (H.4)) are essential in order to get uniqueness.

Proof of theorem 3.1 :

We perform the usual singular change of coordinates

$$X = x, \; t = (\delta - |X|^2)T, \; \delta > 0 \text{ small}.$$

Let us denote by $\tilde{P}_o$ the operator P_o in the new coordinates. We see easily that, setting $|X|^2 = r^2$ and $p = p_1 + ip_2$,

$$Q(X,T;D_X,D_T) = (\delta - |X|^2)^2 \; \tilde{P}_o = \frac{\partial^2}{\partial t^2} +$$

$$+ (\delta - r^2)^2 \; p(X,(\delta - r^2)T;D_X) + \alpha(X,T) D_T$$

where $\alpha(X,T)$ is smooth.

On the other hand, the inequality (3.2) becomes

$$(3.3) \qquad |Q\tilde{u}(X,T)| \leqslant C \left\{ \frac{|\tilde{u}(X,T)|}{T^2} + \frac{|D_T \tilde{u}(X,T)|}{T} \right\} .$$

So we shall work with

$$(3.4) \qquad \begin{cases} L = L(X,T;D_X;D_T) = \dfrac{\partial^2}{\partial t^2} + (\delta - r^2)^2 \, q(X,T;D_X) \\[2ex] q(X,T;D_X) = p(X,(\delta - r^2)T,D_X) . \end{cases}$$

and $L\tilde{u}(X,T)$ is also bounded by the right hand side of (3.3).

Let us now check how conditions (H.i), $1 \leq i \leq 4$ read on q. Denoting $\sigma(q) = q_1 + iq_2$, $\sigma(p) = p_1 + ip_2$, $q_j(X,T,\xi) = p_j(X,(\delta-r^2)T;\xi)$ $j=1,2$, we have by (H.1),

$$\frac{\partial q_2}{\partial T} = (\delta-r^2)\frac{\partial p_2}{\partial t}(X,(\delta-r^2)T;\xi) = (\delta-r^2)\frac{a(X,(\delta-r^2)T)}{(\delta-r^2)T}\, p_2(X,(\delta-r^2)T;\xi)$$

so

$$\text{(H.1)}' \qquad [\frac{\partial}{\partial T}, q_2](X,T;\xi) = \frac{a(X,(\delta-r^2)T)}{T} q_2(X,T;\xi),$$

and in the same manner,

$$\text{(H.2)}' \qquad [\frac{\partial}{\partial t}, q_1] = \frac{\tilde{b}}{T} q_1 + \frac{\tilde{c}}{T} q_2$$

where $\tilde{f}(X,T) = f(X,(\delta-r^2)T)$.

Condition (H.3)' is obtained by taking $\tilde{a},\tilde{b},\tilde{c}$ instead of a,b,c in (H.3).

Finally, using (H.4), (H.1) and (H.2) we see easily that there exist bounded functions f_1,f_2,g_1,g_2 such that

$$\text{(H.4)}' \qquad \{q_1,q_2\} = f_1q_1 + f_2q_2 + \frac{1}{(\delta-r^2)}(g_1q_1 + g_2q_2).$$

In the following we shall use the small letters (x,t) instead of (X,T).

The following Carleman estimate will imply, as usual, the result of theorem 3.1.

<u>Proposition 3.3</u>:

<u>Under conditions</u> (H.i)', $i=1,\dots,4$, <u>there exist positive constants</u> C,T_o,γ_o,r <u>such that for</u> $\gamma \geq \gamma_o$, $T \leq T_o$,

$$(3.5) \qquad \gamma^3 \|t^{-\gamma-2}w\|_o^2 + \gamma\| t^{-\gamma-1}\frac{\partial w}{\partial t}\|_o^2 \leq C\|t^{-\gamma}Lw\|_o^2$$

<u>for all</u> $w \in C^\infty(\mathbb{R}^{n+1})$ <u>with</u> $\operatorname{supp} w \subset \{0 \leq t \leq T_o, |x| \leq r\}$; <u>here</u> $\| \|_o = \| \|_{L^2}$.

Proof :

Let us set $t^{-\gamma}w = v$. It follows that

$$t^{-\gamma}Lw = \tilde{L}v = (\partial_t^2 + \gamma(\gamma-1)t^{-2} + 2\gamma t^{-1}\partial_t + (\delta-r^2)^2 q_1 + $$

$$+ i(\delta-r^2)^2 q_2)v.$$

Setting

$$(3.6) \qquad \begin{cases} A = \partial_t^2 + \gamma(\gamma-1)t^{-2} + (\delta-r^2)^2 q_1 \\ \\ B = 2\gamma t^{-1}\partial_t + i(\delta-r^2)^2 q_2 \end{cases}$$

we get

$$(3.7) \qquad \|\tilde{L}v\|^2 = \|Av\|^2 + \|Bv\|^2 + 2\mathrm{Re}(Av,Bv).$$

It is easy to see that

$$(3.8) \qquad 2\mathrm{Re}(\partial_t^2 v, 2\gamma t^{-1}\partial_t v) = 2\gamma\|t^{-1}\partial_t v\|^2$$

$$(3.9) \qquad 2\mathrm{Re}(\gamma(\gamma-1)t^{-2}v, 2\gamma t^{-1}\partial_t v) = 6\gamma^2(\gamma-1)\ \|t^{-2}v\|^2.$$

Now

$$(\partial_t^2 v, (\delta-r^2)^2 iq_2 v) = -\underbrace{(\partial_t v, (\delta-r^2)^2 i[\partial_t, q_2]v)}_{①} -$$

$$- \underbrace{(\partial_t v, i(\delta-r^2)^2 iq_2\ \partial_t v)}_{②}.$$

Using (H.1)' and (3.6) we see that

$$2\mathrm{Re}\ ① = 4\gamma(\tfrac{1}{t}\partial_t v, a.\tfrac{1}{t}\partial_t v) - 2\mathrm{Re}(\tfrac{1}{t}\partial_t v, a\ Bv).$$

On the other hand,

$$② = (\delta_t v, [(\delta-r^2)^2, q_2]\, i\partial_t v) + (q_2^* \partial_t v, i(\delta-r^2)^2 \partial_t v).$$

Since $q_2^* = q_2 + f(x,t)$ where $f \in C^\infty$ and $[(\delta-r^2)^2, q_2] \in C^\infty$, it follows that

$$2\mathrm{Re}\ ② = (\partial_t v\ ,\ g\partial_t v)$$

where $g \in C^\infty$. Summing up, we have proved

$$(3.10)\quad \begin{cases} 2\mathrm{Re}(\partial_t v, (\delta-r^2)^2\ iq_2 v) = 4\gamma(t^{-1}\partial_t v, at^{-1}\partial_t v) + R_1 \\ \forall\ \varepsilon>0\ \exists C_\varepsilon>0 :\ |R_1| \leqslant \varepsilon\|Bv\|^2 + C_\varepsilon\|t^{-1}\partial_t v\|^2 + C\ \|\partial_t v\|^2. \end{cases}$$

On the other hand,

$$((\delta-r^2)^2 iq_2 v, t^{-2}v) = (i[(\delta-r^2)^2, q_2]v, t^{-2}v) + (i(\delta-r^2)^2 v, t^{-2}q_2^* v).$$

Since $q_2^* = q_2 + f$ we get

$$(3.11)\quad |2\mathrm{Re}((\delta-r^2)^2\ iq_2 v, t^{-2}v)| \leqslant C\ \gamma(\gamma-1)\|\ t^{-1}v\|^2.$$

Let us now check $2\mathrm{Re}((\delta-r^2)^2\ q_1 v,\ 2\gamma t^{-1}\partial_t v)$. We can write

$$((\delta-r^2)^2 q_1 v, t^{-1}\partial_t v) = ([(\delta-r^2)^2, q_1]v, t^{-1}\partial_t v) +$$

$$+ ((\delta-r^2)^2 v, t^{-1}[q_1^*, \partial_t]v) + ((\delta-r^2)^2 v, t^{-2}q_1^* v) -$$

$$- ((\delta-r^2)^2\partial_t v, t^{-1}q_1^* v) = 1/ + 2/ + 3/ - 4/.$$

We have

$$|1/| \leqslant C\gamma\ \|t^{-1}v\|\,.\,\|\partial_t v\|$$

$$2/ = -((\delta-r^2)^2 v,\ \frac{b}{t^2}\, q_1 v + \frac{c}{t^2}\, q_2 v) + ((\delta-r^2)^2 v,\ \frac{\phi}{t}\, v)$$

where b and c are given by (H.2)' and $\phi \in L^\infty$

$$3/ = ((\delta-r^2)^2 v,\ t^{-2}q_1 v) + ((\delta-r^2)^2 v,\ \psi t^{-2}v)\ ,\ \psi \in L^\infty$$

$$4/ = -((\delta-r^2)^2\partial_t v,\ t^{-1}q_1 v)+((\delta-r^2)^2\,\partial_t v,\ t^{-1}\theta v),\ \theta \in L^\infty.$$

It follows that

$$(3.12)\quad 2\mathrm{Re}((\delta-r^2)^2 q_1 v,\ 2\gamma t^{-1}\partial_t v) = -2\gamma((\delta-r^2)^2 v,(b-1)t^{-2}q_1 v)- - 2\gamma((\delta-r^2)^2 v,\ ct^{-2}q_2 v) + R_2$$

with $|R_2| \leqslant C\ \gamma(\|t^{-1}v\|^2 + \|\partial_t v\|^2)$.

Now by (3.6)

$$(\delta-r^2)^2 q_1 v = Av - \partial_t^2 v - \gamma(\gamma-1)t^{-2}v\ ,$$

from which we get

$$- 2\gamma(v,(b-1)t^{-2}(\delta-r^2)^2 q_1 v) = - 2\gamma(t^{-2}v,(b-1)Av) + + 2\gamma(t^{-2}v,(b-1)\partial_t^2 v) + 2\gamma^2(\gamma-1)(t^{-2}v,\ (b-1)t^{-2}v).$$

It follows that

$$(3.13)\quad \begin{aligned} &- 2\gamma(v,(b-1)t^{-2}(\delta-r^2)^2 q_1 v) = -2\gamma(t^{-2}v,(b-1)Av) - \\ &-2\gamma(t^{-1}\partial_t v,(b-1)t^{-2}\partial_t v) + 2\gamma^2(\gamma-1)(t^{-2}v,\ (b-1)t^{-2}v) + R_3 \end{aligned}$$

with $|R_3| \leqslant C\ \gamma\|t^{-2}v\|\cdot\|t^{-1}\partial_t v\|$.

By the same argument we get

$$(3.14)\quad -2\gamma((\delta-r^2)^2 v,\ ct^{-2}q_2 v) = 2\gamma(t^{-2}v,\ icBv) - 4\gamma^2(t^{-2}v, ict^{-1}\partial_t v).$$

It follows from (3.12), (3.13), (3.14) that

$$(3.15)\quad \begin{cases} 2\mathrm{Re}((\delta-r^2)^2 q_1 v,\ 2\gamma t^{-1}\partial_t v) = -2\gamma(t^{-1}\partial_t v,(b-1)t^{-1}\partial_t v) + \\ + 2\gamma^2(\gamma-1)(t^{-2}v,(b-1)t^{-2}v) - 4\gamma^2(t^{-2}v, ict^{-1}\partial_t v) + R_4 \end{cases}$$

$$\forall\, \varepsilon >0 \,\exists\, C_\varepsilon > 0 : |R_4| \leqslant \varepsilon\,((\|Av\|^2 + \|Bv\|^2 + \gamma\|t^{-1}\partial_t v\|^2) + \tag{3.16}$$
$$+ C_\varepsilon \gamma^2 \|t^{-2}v\|^2 + C\gamma\|\partial_t v\|^2 .$$

Finally we consider $2\mathrm{Re}((\delta-r^2)^2 q_1 v,\ (\delta-r^2)^2\ iq_2 v)$. Let us recall that $q_j^* = q_j + f_j$, $f_j \in C^\infty$ $j=1,2$. Setting $\phi=(\delta-r^2)^2$ we can write

$$(\phi q_1 v, \phi i q_2 v) = ([\phi^2, q_1]v, iq_2 v) + (\phi v, i\phi[q_1^*, q_2]v) +$$
$$+([q_2^*, \phi^2]v, iq_1^* v) + (i\phi q_2^* v, \phi q_1^* v).$$

Using (H.4)' it is easy to see that we can find two bounded functions α and β such that

$$2\mathrm{Re}(\phi q_1 v,\ \phi i q_2 v) = \mathrm{Re}(\alpha v,\ \phi q_2 v) + \mathrm{Re}(\beta v,\ \phi q_1 v).$$

It follows then from (3.6) that for every $\varepsilon > 0$ there exists $C_\varepsilon > 0$ such that :

$$|2\mathrm{Re}((\delta-r^2)^2 q_1 v, (\delta-r^2)^2 iq_2 v)| < \varepsilon\,(\|Av\|^2 + \|Bv\|^2) + \tag{3.17}$$
$$+ C_\varepsilon \|v\|^2 + C(\gamma\|\partial_t v\|^2 + \gamma^2\|t^{-1}v\|^2).$$

Using (3.7), (3.8), (3.9), (3.10), (3.15), (3.16), (3.17) we get

$$\left\{\begin{aligned} &\|\tilde{L}v\|^2 = \|Av\|^2 + \|Bv\|^2 + \mathrm{Re}(t^{-1}\partial_t v, Gt^{-1}\partial_t v) + \\ &\quad + \mathrm{Re}(t^{-2}v, Ht^{-2}v) + \mathrm{Re}(it^{-1}\partial_t v, Kt^{-2}v) + R_5 \end{aligned}\right. \tag{3.18}$$

where $G = 2\gamma(2(a+1)-b)$, $H = 2\gamma^2(\gamma-1)(b+2)$, $K = -4\gamma^2 c$ and

$$\left\{\begin{aligned} &|R_5| \leqslant \varepsilon(\|Av\|^2 + \|Bv\|^2 + \gamma\,\|t^{-1}\partial_t v\|^2) + \\ &\quad + C_\varepsilon(\|t^{-1}\partial_t v\|^2 + \gamma\|\partial_t v\|^2 + \gamma^2\|t^{-2}v\|^2). \end{aligned}\right. \tag{3.19}$$

It follows from (H.3)' that

$$\text{(3.20)}\quad \begin{aligned} &\operatorname{Re}(t^{-1}\partial_t v, Gt^{-1}\partial_t v) + \operatorname{Re}(t^{-2}v, Ht^{-2}v) + \\ &+ \operatorname{Re}(it^{-1}\partial_t v, Kt^{-2}v) \geqslant C_o(\gamma\|t^{-1}\partial_t v\|^2 + \gamma^3\|t^{-2}v\|^2). \end{aligned}$$

Using (3.18), (3.19) and (3.20) we get

$$\|\tilde{L}v\|^2 \geqslant C_1(\gamma\|t^{-1}\partial_t v\|^2 + \gamma^3\|t^{-2}v\|^2),$$

which proves Proposition 3.3.

Remark 3.4 :

Let us consider in $\mathbb{R}^2$ the operator $P = \partial_t^2 + a\partial_x$ where $a \in C^\infty(\mathbb{R}^2, \mathbb{R})$. The condition (H.1) says roughly that near $t=0$ the function a behaves like t^k for some k. For instance, theorem 3.1 does not apply to the case $a(x,t) = t-x^2$. But using the result of theorem 1.1 when $a(x,t) = t^k\,\alpha(x,t)$, $\alpha(0,0) \neq 0$ and the argument used in the proof of theorem 4.1 Chapter One we can prove the following

Proposition 3.5 :

Let $P = \partial_t^2 + a(x,t)\partial_x + b\partial_t + c$ *where* a,b *and* c *are* C^∞ *and* a *is real. Let us suppose that*

$t \mapsto a(0,t)$ *has at* $t=0$ *a zero of finite order*

Then there exists a neighborhood W *of the origin in which every* $u \in C^\infty$, *satisfying* $Pu = 0$ *and* $u_{|t<0} = 0$, *must vanish*.

We are going to see what happens if one of the conditions (H.i) i=1,2,3 is not satisfied (in a generic sense). Condition (H.1) means mainly that the linear forms $[\frac{\partial}{\partial t}, p_2]$ and p_2 are dependant for every $(x,t) \in V_+$. The case where they are independant will be separated into two cases. Let us set

$$E = \{q = \sum_{j=1}^{n} a_j(x,t)\xi_j \;:\; a_j \in C^\infty(V, \mathbb{R})\}.$$

Case 1 :

We shall consider here the case where $[\partial_t, p_2]$ and p_2 are independant, while p_1 and p_2 are dependant, which means that $p_1 = \alpha p_2$ (or $p_2 = \alpha p_1$) for some function α. So $p_1 + ip_2 = (\alpha+i)p_2$ (or $(1+i\alpha)p_1$). We can state the

Theorem 3.6 :

Let q be in E and c a non-vanishing C^∞ function in V. We consider the operator $P = \partial_t^2 + cq + \alpha\partial_t + \beta$ where α, β are in $C^\infty(V)$. Let us suppose that

(3.21) *there exists $\xi^o \in \mathbb{R}^n$ such that $q(0,0;\xi^o) = 0$ but $[\partial_t, q](0,0;\xi^o) \neq 0$.*

Then there exist two C^∞ functions a and u in V, vanishing for $t < 0$ such that

$$
(3.22)\qquad \begin{cases} Pu + au = 0 \\ (0,0) \in \operatorname{supp} u . \end{cases}
$$

The simplest example for which theorem 3.1 applies is the following operator in $\mathbb{R}^3$

$$
P = \frac{\partial^2}{\partial t^2} + \frac{\partial}{\partial x_1} + t\frac{\partial}{\partial x_2} .
$$

Case 2 :

Here $[\partial_t, p_2]$ and p_2, p_1 and p_2 are independant.

Theorem 3.7 :

Let $p = \partial_t^2 + p_1 + ip_2 + \alpha\partial_t + \beta$. *Let us suppose that*

(3.23) *there exists* $\xi^0 \in \mathbb{R}^n$ *such that* $p_2(0,0;\xi^0)=0$ *but* $p_1(0,0;\xi^0)\neq 0$, $[\partial_t, p_2](0,0;\xi^0)\neq 0$.

Then the same conclusion as in theorem 3.6 *holds*.

Example 3.8 :

$$P = \partial_t^2 + D_1 + i(D_2 + tD_1)$$

$$P = \partial_t^2 + D_1 + i(D_2 + tD_3).$$

We suppose now (H.1) satisfied but not (H.2). On the other hand, we shall assume that the coefficients of p_j are independant of x.

Theorem 3.9 :

Let $P = \partial_t^2 + p_1 + ip_2 + \alpha\partial_t + \beta$ *where* $p_j = p_j(t;\xi)$, $j=1,2$. *Assume that*

(i) *There exists* d *such that* $[\partial_t, p_2] = dp_2$ *for* $t > 0$.

(ii) *The linear forms* $\frac{\partial p_1}{\partial t}(0;\xi)$, $p_1(0;\xi)$ *and* $p_2(0;\xi)$ *are independant*.

Then the same conclusion as in theorem 3.6 *holds*.

Example $\quad P = \partial_t^2 + D_2 + tD_3 + iD_1.$

The last result will concern the case when (H.1) and (H.2) are satisfied but (H.3) is not.

Theorem 3.10 :

Let P be one of the following operators :

(1) $P = \frac{\partial^2}{\partial t^2} + t^b D_1 + it^a D_2$; $a, b \in \mathbb{N}$ and $2(a+1) < b$

(2) $P = \frac{\partial^2}{\partial t^2} + t^\ell (D_1 + dt^m D_2) + it^{\ell+m} D_2$; $\ell, m \in \mathbb{N}$, $d \in \mathbb{R}$ and

$m^2 d^2 > (\ell+2) \cdot (2+2m+\ell)$.

Then the same conclusion as in theorem 3.6 holds for P.

The reader will compare theorem 3.10 with examples 3.2 c) and d).

Proof of Theorems 3.6 to 3.10 :

All these theorems will be proved along the same lines by the method used in the proof of theorem 3.1 Chapter One. The main point of that proof is the existence of the "phase functions" ξ and ϕ with good convexity properties as shows the following

Proposition 3.11 :

Let $P = \partial_t^2 + p_1 + ip_2 + \alpha\partial_t + \beta$. Let us suppose that there exist a neighborhood ω of the origin in $\mathbb{R}^n_x$, $\theta \in \mathbb{N}$, $\delta_o > 0$, $s_o > 0$, a real function $\xi(x,\delta)$ C^∞ in $\omega \times]0, \delta_o]$, a function $\phi(x,s,\delta)$ C^∞ in $\omega \times [-s_o, s_o] \times [0, \delta_o]$ such that

(i) $(\frac{\partial\phi}{\partial s})^2 + p(x, \delta + s\delta^\theta, \nabla_x \xi(x,\delta)) = 0$

(ii) $\exists\ M \in \mathbb{N}$: $|\xi(x,\delta)| \leq C\ \delta^{-M}$ for $(x,\delta) \in \omega \times [0,\delta_0]$

(iii) $\frac{\partial\phi}{\partial s}(0,0,0) \neq 0$ and $\mathrm{Re}\phi = \alpha(x,\delta)s - \beta(x,s,\delta)s^2$

and we have one of the two following situations

case a) $\theta > 2$, $\alpha(0,0) > 0$, $\beta(0,0,0) > 0$

case b) $\alpha(x,\delta) \equiv 0$ and there exist $\mu \in \mathbb{N}$, $\beta_0 \in C^\infty(\omega)$, $\beta_0(0) > 0$ and $C \geq 0$ such that
$|\beta(x,s,\delta) - \beta_0(x)\delta^\mu| \leq C\ \delta^{\mu+1}$ for $|s| \leq \delta^{\mu+1}$ and $(x,\delta) \in \omega \times [0,\delta_0]$.

Then we can find a neighborhood V of the origin in $\mathbb{R}^{n+1}$, two functions a and u, C^∞ in V, vanishing for $t < 0$, such that

$$\begin{cases} Pu + au = 0 \\ (0,0) \in \mathrm{supp}\, u \ . \end{cases}$$

Before sketching the proof of this result, let us show how it implies theorem 3.6 to 3.9 (The proof of theorem 3.10 will be given later).

Proof of Theorem 3.6 :

In that case the phase equation (i) is

$$(3.24) \qquad \left(\frac{\partial\phi}{\partial s}\right)^2 + c(x,\delta+\delta^\theta s)\ q(x,\delta+\delta^\theta s;\nabla_x\xi(x,\delta)) = 0\ .$$

We can assume without loss of generality that the coefficient of $\frac{\partial}{\partial x_1}$ in q does not vanish in V. Let $\tilde{\xi}(x,\delta)$ be a smooth function for $(x,\delta) \in \omega \times [0,\delta_0]$ solution of

$$(3.25)\quad \begin{cases} q(x,\delta;\nabla_x\tilde{\xi}(x,\delta)) = \varepsilon\,\delta^{\theta}, & (\varepsilon = \pm 1 \text{ to be chosen}) \\ \tilde{\xi}_{|x_1=0} = \sum_{j=2}^{n} x_j\,\xi_j^o \end{cases}$$

where ξ^o is given by (3.21).

It is easy to see that $\nabla_x\tilde{\xi}(0,0) = \xi^o$. Let us set

$$\xi(x,\delta) = \delta^{-\theta}\tilde{\xi}(x,\delta) \qquad , B(x,\delta) = \frac{\partial q}{\partial t}(x,\delta;\nabla_x\tilde{\xi}).$$

It follows from (3.21) that

$$(3.26)\qquad B(0,0) \neq 0.$$

Using Taylor's formula in (3.24) and (3.25) we get

$$(\frac{\partial\phi}{\partial s})^2 + \varepsilon\,c(x,\delta) + s\,\{cB + \frac{\partial c}{\partial t}\,\varepsilon\,\delta^{\theta}\} +$$
$$+ s^2\,\delta^{2\theta}\int_0^1 (1-u)\,\frac{\partial^2}{\partial t^2}\,(cq)(\delta+us\delta^{\theta};\nabla_x\xi)du = 0,$$

which can be written

$$(3.27)\qquad (\frac{\partial\phi}{\partial s})^2 + \varepsilon\,c + s\{cB + \delta f\} + s^2 g = 0$$

where f and g are C^{∞} in (x,δ) with bounded derivatives when $\delta \to 0$.

Since $c(0,0) \neq 0$ we can choose $\varepsilon = \pm 1$ in order to have $-\varepsilon c \notin \mathbb{R}^-$. For (x,s,δ) near the origin we take as ϕ the solution vanishing at $s = 0$ of

$$(3.28)\qquad \frac{\partial\phi}{\partial s} = (-\varepsilon c - s(cB+\delta f) - s^2 g)^{1/2},$$

the square root having positive real part.

Now by (3.28) $\frac{\partial\phi}{\partial s}\big|_{s=0} = (-\varepsilon c)^{1/2}$, and taking the derivative of (3.27) we get $\frac{\partial^2\phi}{\partial s^2}(0,0,0) = \frac{-B(0,0)\ c(0,0)}{2\frac{\partial\phi}{\partial s}(0,0,0)}$ which is $\neq 0$ by (3.26). We cannot have $\mathrm{Re}\ \frac{c(0,0)}{\frac{\partial\phi}{\partial s}(0,0,0)} = 0$ since then $\frac{\partial\phi}{\partial s}(0,0,0) \in i\mathbb{R}$. Taking $-\xi^o$ instead of ξ^o, we can choose the sign of $B(0,0)$ in order to have $\mathrm{Re}\frac{\partial^2\phi}{\partial s^2}(0,0,0) < 0$.

All the conditions of the case a) of Proposition 3.11 are then satisfied.

<u>Proof of theorem 3.7</u> :

We assume that the coefficient of $\frac{\partial}{\partial x_1}$ in p_2 does not vanish in V.

We can then find $\xi(x,\delta)$ C^∞ in $\omega \times [0,\delta_o[$ such that, with ξ^o given by (1.23),

$$(3.29)\qquad \begin{cases} p_2(x,\delta;\nabla_x\xi(x,\delta)) = 0 \\ \xi\big|_{x_1=0} = \sum_{j=2}^{n} x_j\ \xi_j^o . \end{cases}$$

It follows that $\nabla_x\xi(0,0) = \xi^o$. Setting $A=p_1(x,\delta;\nabla_x\xi)$, $B = \frac{\partial p_2}{\partial t}(x,\delta;\nabla_x\xi)$, we deduce from (3.23) that $A(0,0) \neq 0$, $B(0,0) \neq 0$.

The phase equation becomes

$$(3.30)\qquad \begin{aligned} &\left(\frac{\partial\phi}{\partial s}\right)^2 + A + s\delta^\theta\left[\frac{\partial p_1}{\partial t}(x,\delta;\nabla_x\xi) + iB\right] + \\ &+ s^2\ \delta^{2\theta} \int_0^1 (1-u)\ \frac{\partial^2 p}{\partial t^2}(\delta+us\delta^\theta;\nabla_x\xi)\,du=0 . \end{aligned}$$

We can assume without loss of generality (taking in (1.23) $-\xi^o$ instead of ξ^o) that $A(0,0) = 1$.

We take, as ϕ, the solution, vanishing for $s=0$, of

$$\frac{\partial\phi}{\partial s} = i\varepsilon\,(A + is\delta^\theta B + s\tilde{b}(x,\delta) + s^2 c(x,s,\delta))^{1/2}$$

where $\tilde{b}$ and c are smooth in $\omega \times [-s_o,s_o]\times[0,\delta_o]$ and $\varepsilon = \pm 1$ is to be chosen.

It follows that $\frac{\partial\phi}{\partial s}\big|_{s=0} = i\varepsilon\sqrt{A}$, and taking the derivative of (3.30), we get

$$\frac{\partial^2\phi}{\partial s^2}\Big|_{s=0} = -\frac{\delta^\theta B}{2\varepsilon\sqrt{A}} - \frac{\tilde{b}}{2i\varepsilon\sqrt{A}}.$$ So we have

$$\mathrm{Re}\phi(x,s,\delta) = s^2\left(-\frac{\delta^\theta B}{2\varepsilon\sqrt{A}} + O(s)\right).$$

The conditions of Proposition 3.11 case b) are then satisfied with

$$\beta_o(x) = \frac{B(0,x)}{4\varepsilon\sqrt{A}(0,x)} \quad \text{and} \quad \varepsilon = \text{sign } B(0,0).$$

<u>Proof of theorem 3.9</u> :

We are going to construct a real function $\xi = \xi(x,\delta)$ such that

$$(3.31) \qquad \begin{cases} p_1(\delta;\nabla_x\xi) = -1 \\ p_2(\delta;\nabla_x\xi) = 0 \\ \frac{\partial p_1}{\partial t}(\delta;\nabla_x\xi) = \delta^{-\theta}B(\delta),\ B(0) = 1. \end{cases}$$

We set $\xi(x,\delta) = \delta^{-\theta}\eta(x,\delta) + \zeta(x,\delta)$ where

$$(3.32)\quad \begin{cases} p_1(\delta;\nabla_x\eta) = 0 \\ p_2(\delta;\nabla_x\eta) = 0 \\ \dfrac{\partial p_1}{\partial t}(\delta;\nabla_x\eta) = 1 \end{cases}$$

$$(3.33)\quad \begin{cases} p_1(\delta;\nabla_x\zeta) = 1 \\ p_2(\delta;\nabla_x\zeta) = 0. \end{cases}$$

We take $\eta = \sum_{i=1}^{n} \eta_i(\delta)x_i$, $\zeta = \sum_{i=1}^{n} \zeta_i(\delta)x_i$ and we solve the linear systems (3.32) and (3.33) using the condition (ii) of theorem 3.9.

By condition (i), $\frac{\partial p_2}{\partial t}(\delta;\nabla\xi) = 0$. On the other hand, it is easily seen that $B(\delta)=1 + \delta^{\theta}\frac{\partial p_1}{\partial t}(x;\nabla_x\xi)$. The phase equation becomes

$$(3.34)\quad (\frac{\partial\phi}{\partial s})^2 -1+sB + s^2\,\delta^{2\theta}\int_0^1 (1-u)\,\frac{\partial^2 p}{\partial t^2}(\delta+us\delta^{\theta};\nabla_x\xi)\,du=0.$$

It is now easy to see that we can find a solution of (3.34) vanishing at $s=0$ and satisfying all the conditions of case a) in Proposition 3.11.

Let us sketch the proof of Proposition 3.11. For motivations and details we refer the reader to the proof of Theorem 3.1 Chapter One. First we give a value to the parameters. We set

case a) $\delta=b_k = k^{-\rho}$, $\rho = \frac{1}{2(\theta-1)}$; $\nu_k=k^{\sigma}$, $\sigma=2$, $\lambda=b_k^{-\theta}$, $\theta \geqslant 2$

and $\tau_k = b_k^{-2\theta} k^{2\sigma}$.

case b) $\delta = b_k = k^{-\rho}$; $\rho > 0$ such that $\rho(\theta+\mu) < 1$;
$\nu_k = k^{\sigma}$, $\sigma \in \mathbb{N}$ such that $\sigma > 2+(\mu+2-2\theta)\rho$, $\sigma > 1+\rho(\mu+2)$;
$\tau_k = b_k^{-2\theta} k^{2\sigma}$.

We set, with ξ and ϕ given in Proposition 3.11,

$$(3.35)\quad v_k(x,t) = e^{i\tau_k \xi(x,b_k)} e^{\nu_k \phi(x,\frac{t-b_k}{b_k^{\theta}},b_k)} z_k(x,\frac{t-b_k}{b_k^{\theta}})$$

for $(x,t) \in \omega \times [b_{k+1}, b_{k-1}]$.

Setting

$$G_k(x,t) = \nu_k \mathrm{Re}\,\phi(x,\frac{t-b_k}{b_k^{\theta}},b_k) - \nu_{k+1} \mathrm{Re}\,\phi(x,\frac{t-b_{k+1}}{b_{k+1}^{\theta}}, b_{k+1})$$

and $I_k(x) = G_k(x,m_k)$ with $m_k = \frac{1}{3} b_k + \frac{2}{3} b_{k+1}$, we prove

case a) $I_k(x) \sim -\alpha(0,x)\, k^{\sigma-1/2}$

case b) $I_k(x) \sim -\frac{1}{3} \beta_o(x) \rho^2 k^{\alpha+\rho(2\theta-2\mu-2)-2}$.

Then we set

$$(3.36)\quad \gamma_k(x) = -\sum_{j=k_o}^{k-1} I_j(x) \begin{cases} = \text{case a)}\ \dfrac{\rho\,\alpha(0,x)}{\alpha+\frac{1}{2}} k^{\sigma+\frac{1}{2}} \\[2ex] = \text{case b)}\ \dfrac{\beta_o(x)\rho^2}{3(\sigma+\rho(2\theta-\mu-2)-1)} k^{\alpha+\rho(2\theta-\mu-2)-1} \end{cases}$$

Let us set $w_j = e^{-\gamma}\,\delta^{(2\theta-d)}\tilde{w}_j$ (where $d = \frac{\sigma+1/2}{\rho}$ in case a) and $d = \frac{\sigma-1}{\rho} + 2\theta-\mu-2$ in case b)), $\gamma(x,\delta) = \delta^{-d}\,\tilde{\gamma}(x,\delta)$, the functions $\tilde{w}_j$ being solutions of the equations

$$2\,\frac{\partial\phi}{\partial s}\cdot\frac{\partial\tilde{w}_o}{\partial s} + \left[\frac{\partial^2\phi}{\partial s^2} + \frac{1}{i\lambda^2}\,p(x,\delta+\frac{s}{\lambda};\nabla_x\phi)\right]\tilde{w}_o = 0$$

$$2\,\frac{\partial\phi}{\partial s}\cdot\frac{\partial\tilde{w}_j}{\partial s} + \left[\frac{\partial^2\phi}{\partial s^2} + \frac{1}{i\lambda^2}\,p(x,\delta+\frac{s}{\lambda};\nabla_x\phi)\right]\tilde{w}_j = -c\delta^d\,\tilde{w}_{j-1} +$$

$$+\, i\delta^d\,p(x,\delta+\frac{s}{\lambda};\nabla_x\tilde{w}_{j-1}) - \delta^{d-2\theta}\,\frac{\partial^2\tilde{w}_{j-1}}{\partial s^2} - ip(x,\delta+\frac{s}{\lambda};\nabla_x\tilde{\gamma}).\tilde{w}_{j-1}.$$

with $\tilde{w}_o|_{s=0} = 1$, $\tilde{w}_j|_{s=0} = 0$.

These equations are not singular when $\delta \to 0$ since $d-2\theta > 0$; they have smooth solutions $\tilde{w}_j$.
We can find $w = w(x,s,\delta,\xi)$ such that as $|\xi| \to 0$,

$$|D_x^\alpha\,D_s^\ell(w - \sum_{j=0}^{N-1}\tilde{w}_j\,\xi^j)| \leq c_N\,|\xi|^N.$$

The function $z_k(x,s)$ in (3.35) is then defined by

$$z_k(x,s) = w(x,s,b_k,\frac{b_k^{2\theta-d}}{\nu_k}).e^{-\gamma_k(x)}.$$

We can prove then that, setting $\tilde{r}_k = \frac{Pv_k}{v_k}$, we have as $k\to\infty$,

(3.37) $$|D^\beta_{(x,t)}\tilde{r}_k| \leq C_{\alpha N}k^{-N}.$$

Let us now set $F_k(x,t) = \mathrm{Log}\,|\frac{v_k}{v_{k+1}}(x,t)|$. Then for $(x,t)\in\omega\times[b_{k+1},b_k]$

$$\frac{\partial F_k}{\partial t} \geqslant C k^{\eta} \tag{3.38}$$

with $C > 0$ and $\eta = \sigma + \rho$ in case a), $\eta = \sigma + \rho(2\theta - \mu - 1) - 1$ in case b).

Using the inequality we prove that we can find a smooth function $m_k(x)$ such that

$$|m_k(x) - m_k| \leqslant C\, k^{-\eta}$$

$$F_k(x, m_k(x)) = 0 .$$

Then, using Whitney's theorem we modify slightly v_k into u_k in such a manner that $r_k = \frac{Pu_k}{u_k}$ is flat on the curves $t = m_k(x)$ and $t = m_{k-1}(x)$.

We then end the proof as in theorem 3.1 Chapter One.

<u>Proof of theorem 3.10</u> :

We deal here with the operator

$$P = \frac{\partial^2}{\partial t^2} + t^{\ell}(D_1 + dt^m D_2) + it^{\ell+m} D_2$$

$\ell \in \mathbb{N}$, $m \in \mathbb{Z}$ ($d=0$, $m<0$ in case (2)). The solutions will be of the form

$$v(x,t) = e^{i\tau\xi(x,\delta)} e^{\nu\phi(t/\delta)} w(t/\delta)$$

where $\xi(x,\delta) = \varepsilon\delta^{-\ell} x_1 + \alpha\delta^{-m-\ell} x_2$, $\varepsilon = \pm 1$ and $\alpha \in \mathbb{R}^*$ is to be chosen. We see easily that

$$e^{-i\tau\xi}\, e^{-\nu\phi} Pv = \nu^2\delta^{-2}\phi'^2 w + \nu\delta^{-2}\phi'' w + 2\nu\delta^{-2}\phi' w' +$$

$$+ \delta^2 w'' + \varepsilon\tau (\tfrac{t}{\delta})^{\ell} w + \alpha(d+i)(\tfrac{t}{\delta})^{m+\ell}\tau w .$$

We set $\tau=\nu^2\delta^{-2}$ and $s = \frac{t}{\delta}$; the phase ϕ will be a solution of

$$(3.39) \qquad \phi'^2 + \varepsilon s^{\ell} + \alpha(d+i)s^{\ell+m} = 0 .$$

We shall take $w(s,\nu) \sim w_o(s) + \frac{1}{\nu} w_1 + \ldots$, the w_j satisfying the transport equations $2\phi' \frac{\partial w_j}{\partial s} + \phi'' w_j = - \frac{\partial^2 w_{j-1}}{\partial s^2}$ $j \geqslant 0$

with $w_j = 0$ if $j<0$, $w_o(1)=1$, $w_j(1)=0$, $j \geqslant 1$.

Let us now solve the phase equation (3.39).
In the first case, $d=0$ and $2(m+\ell+1) < \ell$, we shall take $\varepsilon=1$ and we solve

$$(3.40) \qquad \begin{cases} \frac{\partial\phi}{\partial s} = i s^{\ell/2}(1+i\alpha s^m)^{\frac{1}{2}} \\ \phi(1) = 0 . \end{cases}$$

In the second one we take $\varepsilon=-1$ and solve

$$(3.41) \qquad \begin{cases} \frac{\partial\phi}{\partial s} = s^{\ell/2}(1-\alpha(d+i)s^m)^{\frac{1}{2}} \\ \phi(1) = 0 . \end{cases}$$

Lemma 3.12 :
With the hypotheses of theorem 3.10, we can find $\alpha \in \mathbb{R}^*$ *such that*

$$(3.42) \qquad \mathrm{Re}\phi'_{|s=1} > 0 \quad \text{and} \quad -\mathrm{Re}\phi''_{|s=1} > \mathrm{Re}\phi'_{|s=1} .$$

Proof :
Case (1) : Let $z = A+iB$ be the square root of $(1+i\alpha)$ with positive real part. It follows from (3.40) that $\phi'_{|s=1} = iz$ so $\mathrm{Re}\,\phi'_{|s=1} = -B$ and $\phi''_{|s=1} = \frac{m}{2iz} - \frac{(m+\ell)z}{2i}$ so

$\mathrm{Re}\phi''|_{s=1} = \frac{-B}{2|z|^2}(m+(m+\ell)|z|^2)$. Since $|z|^2 = \sqrt{1+\alpha^2}$, the second inequality of (3.42) can be written, if $B<0$, $-m > (m+\ell+2)\sqrt{1+\alpha^2}$. This inequality is satisfied for α small enough, since by hypothesis $2m+\ell+2<0 \iff m+\ell+2<-m$.

<u>Case (2)</u> : Let $z = A+iB$ the square root of $1-\alpha(d+i)$ with positive real part. We deduce from (3.41), $\phi'|_{s=1} = z$ and $\phi''|_{s=1} = \frac{m+\ell}{2}z - \frac{m}{2z}$, so $\mathrm{Re}\phi''|_{s=1} = \frac{A}{2|z|^2}(-m+(m+\ell)|z|^2)$. Since $A>0$ the first inequality in (3.42) is satisfied and the second one too if $m-(m+\ell+2)|z|^2>0$, which is equivalent to $\alpha(d^2+1)-2\alpha d+1-\frac{m^2}{(m+\ell+2)^2}<0$. Our hypothesis implies that the discriminant $\Delta = d^2-(d^2+1)(1-\frac{m^2}{(m+\ell+2)^2})>0$.

So we can find $\alpha \neq 0$. Let us now sketch the construction of u. We shall take

$\delta = b_k = k^{-\rho}$, $\nu = \nu_k = k^{\sigma}$, $\tau_k = \nu_k^2 b_k^{-2}$ (ρ,σ to be chosen) and set

$$v_k(x,t) = e^{-\gamma_k} e^{i\tau_k \xi(x,b_k)} e^{\nu_k\phi(t/b_k)} w(t/b_k,\nu_k).$$

Setting $F_k = \mathrm{Log}\left|\frac{v_k}{v_{k+1}}\right|$, we then have :

a) If $-(\sigma+\rho)\,\mathrm{Re}\phi'(1) - \rho\,\mathrm{Re}\phi''(1) > 0$ and $\sigma>2$ then

$$\frac{\partial F_k}{\partial t} \geqslant c\, k^{\sigma+\rho-1} \geqslant c\, k^{\rho+1}.$$

b) Let $m_k = \frac{1}{2}(b_k+b_{k+1})$; we can find γ_k such that

$F_k(m_k) = 0$. Moreover, $\gamma_k \sim \frac{\rho}{\sigma}\mathrm{Re}\phi'(1)k^{\sigma}$ when $k\to\infty$.

In order to get a) and b) it is sufficient, by Lemma 3.12, to take $\sigma=3$ and ρ big enough such that

$-\mathrm{Re}\phi''(1) > \frac{3+\rho}{\rho}\mathrm{Re}\phi'(1)$. We end the proof as in Theorem 3.1 Chapter one. ∎

§ 4. ELLIPTIC OPERATORS

We shall deal in this section with an elliptic operator, for which we shall give results of uniqueness and non-uniqueness. Let us first recall that the hypersurface $S = \{\varphi(x) = \varphi(x_o)\}$ is called strongly pseudo-convex with respect to such an operator P (with principal symbol p) at x_o, if "For all $\zeta = \xi + i\tau \cdot \operatorname{grad} \varphi(x_o)$, $\xi \in \mathbb{R}^n$, $\tau \neq 0$, satisfying $p(x_o,\zeta) = \{p,\varphi\}(x_o,\zeta) = 0$, we have

$$\sum_{j,k=1}^{n} \frac{\partial^2\varphi}{\partial x_j \partial x_k}(x_o)\, \overline{\frac{\partial p}{\partial \xi_j}(x_o,\zeta)}\, \frac{\partial p}{\partial \xi_k}(x_o,\zeta) +$$

$$+ \frac{1}{\tau} \operatorname{Im} \sum_{k=1}^{n} \frac{\partial p}{\partial x_k}(x_o,\zeta)\, \overline{\frac{\partial p}{\partial \xi_k}(x_o,\zeta)} > 0 \text{ ".}$$

We can now state the

Theorem 4.1 :

Let $P(x,D)$ be an elliptic differential operator of order m, defined in a neighborhood V of a point x_o, with C^2 principal part and bounded lower order terms. Let $S = \{x \in V : \varphi(x) = \varphi(x_o)\}$ be a C^2 hypersurface which is strongly pseudo-convex with respect to P at x_o. Then there exists a neighborhood W of x_o such that if $u \in C^m(V)$ satisfies $P(x,D)\, u = 0$ and $u=0$ in $\{x \in V : \varphi(x) > \varphi(x_o)\}$ then $u = 0$ in W.

As usual, the uniqueness will be a consequence of a Carleman estimate which we state now.

Theorem 4.2 :

Let P and S be as in theorem 4.1. We can then find a neighborhood Ω of x_o and a positive constant C such that for every $v \in C_o^\infty(\Omega)$ and every $\gamma \geq e^C$, we have

$$\text{(4.1)} \qquad \sum_{|\alpha| \leq m} \gamma^{m-|\alpha| - \frac{1}{2}} \, \| D^\alpha v \|_{L^2} \leq C \, \| P_\gamma v \|_{L^2}$$

where $P_\gamma = e^{\gamma\psi}\, P\, e^{-\gamma\psi}$ *and*

$$\psi(x) = \varphi(x_o) + \varphi'(x_o).(x-x_o) + \frac{1}{2}\varphi''(x_o)(x-x_o)^2 + \\ + \frac{C}{2}(\varphi'(x_o)(x-x_o))^2 - \frac{1}{2C^2}\|x-x_o\|^2 .$$

Proof

Let us consider the operator

$$(4.2) \qquad Q_\gamma = \frac{C^3}{2}(P_\gamma^* P_\gamma + P_\gamma P_\gamma^*) + \frac{\gamma}{2}[P_\gamma^*, P_\gamma]$$

where C is a constant and let us set

$$(4.3) \qquad q_C = \text{the Weyl symbol of } Q_\gamma .$$

Then, using (1.8) and (1.9) in §1, we get

$$(4.4) \qquad q(x,\xi,\gamma) = C^3|p(x,\zeta)|^2 + C^3\, 2\mathrm{Re}\,\overline{p(x,\zeta)}.p_{m-1}(x,\zeta) + \\ + C\gamma^2|p'_\xi(x,\zeta).\varphi'(x_o)|^2 + \gamma^2\varphi''(x_o).\overline{p'_\xi(x,\zeta)}.p'_\xi(x,\zeta) + \\ + \gamma\,\mathrm{Im}\,\overline{p'_\xi(x,\zeta)}.p'_x(x,\zeta) - \frac{\gamma^2}{C^2}\overline{p'_\xi(x,\zeta)}.p'_\xi(x,\zeta) + \\ + \sum_{\substack{j\geq 2\\ 0\leq 2\ell\leq j}} \gamma^{2\ell} b_{2m-j-2\ell}(x,\zeta)\Pi_{2\ell}(C).C^3 + \\ + \sum_{\substack{j\geq 2\\ 0\leq 2\ell+1\leq j}} \gamma^{2\ell+2} b_{2m-j-2\ell-1}(x,\zeta)\Pi_{2\ell+1}(C)$$

where $\zeta = \xi + i\gamma\psi'(x)$, $b_j(x,\zeta)$ are homogeneous in ζ of order j and $\lim_{C\to\infty} \Pi_\ell(C).C^{-\ell} = a_\ell \in \mathbb{C}$.

We can now state the key result of the proof.

Lemma 4.3

We can find a constant $C>0$ such that for all $|x-x_o| \leq C^{-2}$, $\gamma \geq e^C$ and $\xi \in \mathbb{R}^n$,

$$(4.5) \qquad q_C(x,\xi,\gamma) \geq C^{-1}|\zeta|^{2m}.$$

Proof.

We use a contradiction argument and say that, otherwise, for every $k \in \mathbb{N}^*$ we can find x_k, γ_k, ξ_k with $|x_k - x_o| \leqslant k^{-2}$, $\gamma_k \geqslant e^k$ and $\xi_k \in \mathbb{R}^n$ such that

$$(4.6) \qquad q_k(x_k,\ \xi_k,\ \gamma_k) < \frac{1}{k}\,|\zeta_k|^{2m}$$

where $\zeta_k = \xi_k + i\psi_k'(x_k)$ and

$$\psi_k(x) = \varphi(x_o) + \varphi'(x_o)(x - x_o) + \frac{1}{2}\,\varphi''(x_o)(x - x_o)^2 + \frac{k}{2}[\,\varphi'(x_o).(x - x_o)]^2 -$$

$$- \frac{1}{2k^2}\,\|x - x_o\|^2.$$

Now since $|x_k - x_o| \leqslant k^{-2}$, $\lim\limits_{k\to\infty} \psi_k'(x_k) = \varphi'(x_o) \neq 0$ so $\zeta_k \neq 0$

nd we can set

$$Z_k = \frac{\zeta_k}{|\zeta_k|}\ ,\quad \Gamma_k = \frac{\gamma_k}{|\zeta_k|}\ ,\quad \Xi_k = \frac{\xi_k}{|\zeta_k|}\ .$$

t follows from (4.4) and (4.6) that

$$(4.7) \quad \underbrace{k^3|p(x_k,Z_k)|^2}_{1/} + \underbrace{\frac{k^3\Gamma_k}{\gamma_k}\,2\mathrm{Re}\,\overline{p(x_k,Z_k)}\,p_{m-1}(x_k,Z_k)}_{2/} +$$

$$+ \underbrace{k\Gamma_k^2\,|p_\xi'(x_k,Z_k)\varphi'(x_o)|^2}_{3/} + \underbrace{\Gamma_k^2\,\varphi''(x_o)\,\overline{p_\xi'(x_k,Z_k)}.p_\xi'(x_k,Z_k)}_{4/} +$$

$$+ \underbrace{\Gamma_k\,\mathrm{Im}\,p_\xi'(x_k,Z_k).p_x'(x_k,Z_k)}_{5/} - \underbrace{\frac{\Gamma_k}{k^2}\,p_\xi'(x_k,Z_k).p_\xi'(x_k,Z_k)}_{6/} +$$

$$+ \underbrace{\sum_{\substack{j\geqslant 2\\ 0\leqslant 2\ell\leqslant j}} \Pi_{2\ell}(k).k^3\,\Gamma_k^{j+2\ell}\,b_{2m-j-2\ell}(x_k,Z_k)\,.\,\gamma_k^{-j}}_{7/} +$$

$$+ \underbrace{\sum_{\substack{j\geqslant 2\\ 0\leqslant 2\ell+1\leqslant j}} \Pi_{2\ell+1}(k)\,\Gamma_k^{j+2\ell+1}\,b_{2m-j-2\ell-1}(x_k,Z_k)\gamma_k^{-j+1}}_{8/} < \frac{1}{k}\ .$$

Taking subsequences, we can assume that when $k \to +\infty$,

$$\Gamma_k \to \Gamma_o, Z_k \to Z_o, \Xi_k \to \Xi_o .$$

Now we see easily that when $k \to +\infty$,

(4.8) $\qquad 2/ \to 0 \; , \; 6/ \to 0 \; , \; 7/ \to 0 \; , \; 8/ \to 0 \; .$

Dividing both sides of (4.7) by k^3 we get

$$|p(x_k, Z_k)|^2 + \varepsilon_1(k) < k^{-4}$$

with $\varepsilon_1(k) \to 0$. It follows that

(4.9) $$p(x_o, Z_o) = 0 .$$

Since $Z_o = \Xi_o + i\Gamma_o \varphi'(x_o)$ and P is elliptic, we have $\Gamma_o \neq 0$. Dividing (4.7) by k and using the positivity of 1/, we get

$$\Gamma_k^2 \, |p'_\xi(x_k, Z_k) . \varphi'(x_o)|^2 + \varepsilon_2(k) < k^{-2} .$$

It follows that

(4.10) $$\sum_{j=1}^{n} \frac{\partial p}{\partial \xi_j}(x_o, Z_o) . \frac{\partial \varphi}{\partial x_j}(x_o) = \{p, \varphi\}(x_o, Z_o) = 0 .$$

Using once more (4.7) and the positivity of 1/ and 3/, we get $4/ + 5/ + \varepsilon_3(k) < k^{-1}$

where $\varepsilon_3(k) \to 0$. Since $\Gamma_o^2 > 0$ we get

(4.11) $$\sum_{i,j=1}^{n} \frac{\partial^2 \varphi}{\partial x_i \partial x_j}(x_o) \overline{\frac{\partial p}{\partial \xi_j}(x_o, Z_o)} . \frac{\partial p}{\partial \xi_j}(x_o, Z_o) + \frac{1}{\Gamma_o} \operatorname{Im} \sum_{j=1}^{n} \frac{\partial p}{\partial \xi_j}(x_o, Z_o) \frac{\partial p}{\partial x_j}(x_o, Z_o) \leq 0 .$$

Then, Lemma 4.3 follows from the fact that (4.9), (4.10) and (4.11) are in contradiction with the strong pseudo-convexity of S. ∎

We deduce from Lemma 4.3 that for $|x - x_o| \leq C^{-2}$, $\gamma \geq e^C$, $\xi \in \mathbb{R}^n$

(4.12) $\sigma(x,\xi,\gamma) = q(x,\xi,\gamma) - \frac{1}{2C}(|\xi|^2+\gamma^2)^m \geq C_o(|\xi|^2 + \gamma^2)^m$.

By (4.12) we can find an approximate square root of the pseudo-differential operator A with Weyl symbol $\sigma(x,\xi,\gamma)$; i.e. we can find B such that

$$A = B^* \# B + R$$

and $|(Ru, u)| \leqq C_1 \sum_{|\alpha| \leq m-1} \gamma^{2(m-1-|\alpha|)} \| D^\alpha u\|_o^2$.

Therefore

(4.13) $$(Au, u) \geq - C_1 \sum_{|\alpha| \leq m-1} \gamma^{2(m-1-|\alpha|)} \| D^\alpha u\|_o^2 .$$

Now

$$A = \frac{C^3}{2}(P_\gamma^* P + P_\gamma P_\gamma^*) + \frac{\gamma}{2} [P_\gamma^*, P_\gamma] - \frac{1}{2C}(\gamma^2-\Delta)^m ,$$

so by (4.13)

$$(\frac{C^3}{2} + \frac{\gamma}{2}) \| P_\gamma u \|_o^2 - (\frac{\gamma}{2} - \frac{C^3}{2}) \|P_\gamma^* u\|_o^2 - \frac{1}{2C}((\gamma^2-\Delta)^m u,u) \geq$$

$$\geq - C_1 \sum_{|\alpha| \leq m-1} \gamma^{2(m-1-|\alpha|)} \|D^\alpha u\|_o^2 ;$$

so we get (4.1). ∎

The importance of the condition of strong pseudo-convexity in order to get uniqueness without any condition on the lower terms is emphasized by the following operator in $\mathbb{R}^3_{(x_1,x_2,t)}$

$$P = (D_t^3 + D_1^2 + D_2^2)^2 + t(D_1^2 + D_2^2)^2 - \frac{1}{4} D_1^2 ,$$

which is elliptic near the origin and for which we can construct a function $a \in C^\infty(\mathbb{R}^3)$ such that $P + a$ does not possess the uniqueness property with respect $\{t = 0\}$. Indeed, denoting by (ξ_1,ξ_2,η) the dual variable of (x_1,x_2,t), the point $\xi_1 = 0$, $\xi_2 = 1$, $\eta = 0$ and $\tau = 1$ is a solution of

$$p(0,0,0; \xi_1, \xi_2,\eta + i\tau) = \frac{\partial p}{\partial \tau}(0,0,0;\xi_1,\xi_2, \eta + i\tau)$$

which does not satisfy the condition of strict pseudo-convexity since the coefficients of P are independant of (x_1,x_2).

The following result, stated without proof, contains the above example. We work in a neighborhood of the origin in $\mathbb{R}^{n+1} = \{(x,t) : x \in \mathbb{R}^n, t \in \mathbb{R}\}$ and the initial hypersurface S is given by $\{(x,t) : t = 0\}$.

Theorem 4.4

Let P be a differential operator of order m with principal symbol $p(x,t;\xi,\tau)$. Let us suppose that there exists $\xi^o \in \mathbb{R}^n\setminus 0$, $\tau^o \in \mathbb{C}\setminus\mathbb{R}$ such that, with $m_o = (0,0;\xi^o,\tau^o)$ we have

(i) $p(m_o) = \frac{\partial p}{\partial \tau}(m_o) = 0$ but $\frac{\partial^2 p}{\partial \tau^2}(m_o) \neq 0$.

(ii) $\ddot{t}(m_o) \neq 0$ (where the dot denotes the derivation along the (complex) bicharacteristic of p (complexified) through m_o).

(iii) The manifold

$$\Sigma = \{(x,\xi) \in T^*\mathbb{R}^{n-1}\setminus 0 : \exists\, \tau \in \mathbb{C} : p(x,0;\xi,\tau) = \frac{\partial p}{\partial \tau}(x,0;\xi,\tau) = 0\}$$

has codimension two and is involutive near $(0,\xi^o)$.

(iv) $\mathrm{Re}(p'_\xi(m_o))$ and $\mathrm{Im}(p'_\xi(m_o))$ are independant.

Then we can find u and a in $C^\infty(\mathbb{R}^{n+1})$ with support in $\{(x,t) : t \geqslant 0\}$ satisfying

$$\begin{cases} Pu + au = 0 \\ (0,0) \in \mathrm{Supp}\, u \end{cases}$$

Let us remark that (iii) can be written

$$\mathrm{Im} \sum_{j=1}^{n} \frac{\partial p}{\partial x_j}(m) \overline{\frac{\partial p}{\partial \xi_j}(m)} = 0$$

and recall that the condition of strong pseudo-convexity is

$$\frac{1}{\tau}\ \mathrm{Im} \sum_{j=1}^{n} \frac{\partial p}{\partial x_j}(m_o)\ \overline{\frac{\partial p}{\partial \xi_j}(m_o)} > 0 .$$

§5. SOME NON-UNIQUENESS RESULTS

We end this chapter by giving, without proofs, several theorems on non-uniqueness.
Let us recall that, given two differential operators with constant coefficients $P(D)$ and $Q(D)$, we say that $Q(D)$ is weaker than $P(D)$ if there exists a constant $C > 0$ such that

$$\sum_{\alpha} |Q^{(\alpha)}(\xi)|^2 \leqslant C \sum_{\alpha} |P^{(\alpha)}(\xi)|^2$$

for every $\xi \in \mathbb{R}^n \setminus 0$.

Theorem 5.1 :

Let $P(D)$ and $Q(D)$ be two differential operators with constant coefficients such that Q is weaker than P. For every $N \in \mathbb{R}^n \setminus 0$ such that $p_m(N) \neq 0$, we can find $a \in C^\infty(\mathbb{R}^n)$, $u \in C^\infty(\mathbb{R}^n)$ vanishing in $\{x : \langle x,N\rangle \leqslant 0\}$ such that

$$\begin{cases} (P(D) + a\,Q(D))u = 0 \\ \mathrm{supp}\, u = \{x : \langle x,N\rangle \geqslant 0\}. \end{cases}$$

This result contains theorem 4.5 Chapter One and can be used to prove the claim after theorem 5.3 Chapter Two.∎

Theorem 5.2 :

Let $P(x,D_x)$ be a differential operator of order m

with C^∞ *coefficients in a neighborhood* V *of the origin in* $\mathbb{R}^n$ *and real principal symbol* $p_m(x,\xi)$. *Assume that*

i) *There exists* $\phi \in C^\infty(V)$, ϕ *real such that*

$p_m(x,\nabla_x\phi(x)) = 0$ *and* $X = \sum_{j=1}^{n} \frac{\partial p_m}{\partial \xi_j}(x,\nabla_x\phi(x))\cdot\frac{\partial}{\partial x_j}$ *does not vanish at the origin.*

ii) *There exists* $\psi \in C^\infty(V)$, ψ *real*, $\psi(0) = 0$, $\nabla_x\psi(0) \neq 0$ *such that for* x *in* V,

$$p_m(x,\nabla\psi(x)) \geq C|\psi(x)|^\alpha$$

for some $C > 0$ *and* $\alpha \in [0,+\infty[$.

iii) *In* V, $X\psi=0$.

Then there exist a neighborhood W *of the origin and functions* $u \in C^\infty(W)$, $a \in C^\infty(W)$ *such that* $\operatorname{supp} u = \{x \in W : \psi(x) \geq 0\}$, $\operatorname{supp} a \subset \{x \in W : \psi(x) \geq 0\}$ *and* $Pu + au = 0$ *in* W.

This result can be applied to the following situation. In $\mathbb{R}^{n+1} = \{(x,t)\ ,\ x \in \mathbb{R}^n, t \in \mathbb{R}\}$ let us consider the operator with C^∞ coefficients

$$P = D_t^2 + a(x,D_x) + \sum_{j=1}^{n} \alpha_j D_{x_j} + \alpha D_t + \gamma$$

where $a(x,\xi)$ is a real quadratic form such that

$$(\star) \qquad \exists\ \xi^o \in \mathbb{R}^n \setminus 0 \ :\ a(0,\xi^o)=0,\ d_\xi a(0,\xi^o) \neq 0.$$

Then the result of theorem 5.2 holds with $\psi(x,t) = t$. For instance, we can take $P = D_t^2 + D_1^2 - D_2^2$ in $\mathbb{R}^3$. ∎

In the following $S = \{\phi(x) = \phi(x_o)\}$ will denote a smooth non-characteristic hypersurface.

Theorem 5.3 :

Let P *be a differential operator with principal symbol* $p(x,\xi) = p_1 + ip_2$. *Let us suppose that there exists* $\eta^o \in \mathbb{R}^n \setminus 0$ *such that*

i) $p(0,\eta^o) = 0$, $\{p,\phi\}\ (0,\eta^o) \neq 0$

ii) dp_1 and dp_2 *are independant at* $(0,\eta^o)$ *and*

$$\Sigma = \{(y,\eta) \in T^\star \mathbb{R}^n \setminus 0 \ : \ p(y,\eta) = 0\}$$

is transverse to the vertical fibers

iii) $\{p_1,p_2\}\ (0,\eta^o) \neq 0$.

Then there exists a neighborhood W *of zero and two* $C^\infty(W)$ *functions* u, a *vanishing in* $\{x : \phi(x) \leqslant \phi(0)\}$ *such that* $Pu + au = 0$ *and* $0 \in \operatorname{supp} u$.

Let us remark that this result contains theorem 3.1 Chapter One.

COMMENTS

1) The notion of strong pseudo-convexity is due to L. Hörmander [19] who also proved Theorem 1.3. The simpler proof given here is due to N. Lerner [29]. By the same method one can also get the uniqueness for principally normal operators (see [19]). For uniqueness results under conditions weaker than strong pseudo-convexity, see L. Nirenberg [36].

The non-uniqueness Theorem 1.8 is taken from S. Alinhac [1]. A systematic study of non-uniqueness for second order operators with real principal part can be found in H. Bahouri [6].

2) The results of §2 are taken from R. Lascar-C.Zuily [27]. The definition of pseudo-convexity is a natural extension of that given in §1 to the quasi-homogeneous symbols. So Theorem 2.3 must be compared with Theorem 1.3, Theorem 2.6 with Theorem 1.8. Related results, when the sub-principal symbol takes complex values, can be found in B. Dehman [15] and in J. Hounie - M.E. Moraes-Melo [22]. See also V.M. Isakov [23].

3) The results of section 3 are due to S. Alinhac-C. Zuily [5] and we refer to that paper for refined results.

4) Theorem 4.1 is taken from L. Hörmander [19], but the proof follows N. Lerner [29] where more general results involving weaker pseudo-convexity conditions can be found. The counterexample after Theorem 4.2 is due to A. Pliš [42]. Theorem 4.4 is taken from S. Alinhac [2] and we refer to that paper for the details of the proof. See also L. Hörmander [20] for counterexamples concerning elliptic operators.

5) Theorem 5.1 follows L. Hörmander [20], Theorem 5.2 is taken from M.S. Baouendi-S.Alinhac [4] and Theorem 5.3 is due to S. Alinhac [1].

REFERENCES

[1] *S. ALINHAC* : Non unicité du problème de Cauchy pour des opérateurs de type principal. Séminaire Goulaouic-Schwartz exposé n° 16. Ecole polytechnique Paris (Mars 1981).

[2] *S. ALINHAC* : Non unicité du problème de Cauchy pour des opérateurs de type principal II. Annals of Math. 117 (1983), 77-108.

[3] *S. ALINHAC* : Unicité du problème de Cauchy pour des opérateurs du second ordre à symbole réel. To appear.

[4] *S. ALINHAC-M.S. BAOUENDI* : Constructions de solutions nulles et singulières pour des opérateurs de type principal. Séminaire Goulaouic-Schwartz exposé n° XXII (1979). Ecole polytechnique. Paris.

[5] *S. ALINHAC-C. ZUILY* : Unicité et non unicité du problème de Cauchy pour des opérateurs hyperboliques à caractèristiques doubles. Comm. on p.d.e. (7) (1981), 799-828.

[6] *H. BAHOURI* : Unicité et non unicité pour des opérateurs à symbole principal réel. Thèse du 3e cycle. Université Paris Sud (Orsay) (1982).

[7] *M.S. BAOUENDI-E.C. ZACHMANOGLOU* : Unique continuation of solutions of partial differential equations and inequalities from manifolds of any dimension. Duke

Math. Journal 45 (1978), 1-13.

[8] *A.P. CALDERON* : Uniqueness of the Cauchy problem for partial differential equations. Amer. Journ. of Math. 80 (1958), 16-36.

[9] *A.P. CALDERON* : Existence and uniqueness theorems for systems of partial differential equations. Proc. Symp. Fluid Dynamics and Applied Math. (Univ. of Maryland 1961). Gordon and Breach New York (1962), 147-195.

[10] *F. CARDOSO-J.HOUNIE* : First order linear pde's and uniqueness of the Cauchy problem. Journ. of differential equations 33 (1979) 239-248.

[11] *T. CARLEMAN* : Sur un problème d'unicité pour les systèmes d'équations aux dérivées partielles à deux variables indépendantes. Ark. Mat. Astr. Fys. 26B n°17 (1939),1-9.

[12] *P. COHEN* : The non-uniqueness of the Cauchy problem. O.N.R. Technical report n°93 Stanford University (1960).

[13] *M. DAMLAKHI-C. ZUILY* : On the uniqueness of the Cauchy problem. Journ. of Differential equations 45, 307-316 (1982).

[14] *E. DE GIORGI* : Un esempio di non unicità della soluzione del problema di Cauchy relativo ad una equazione differenziale lineare a derivate parziali di typo parabolico. Rendic. Di Mat et della sua Applic. ser. 5 Vol 14 (1955), 382-387.

[15] *B. DEHMAN* : Unicité et non unicité du probleme de Cauchy pour une classe d'opérateurs quasi-homogènes. Thèse de 3e cycle. Université Paris-Sud (Orsay)(1982).

[16] *P.M. GOORJAN* : The uniqueness of the Cauchy problem for partial differential equations which may have multiple characteristics. Trans. Amer. Math. Soc. 146 (1969), 493-509.

[17] *L. HÖRMANDER* : On the uniqueness of the Cauchy problem. Math. Scand. 6 (1958), 213-225.

[18] *L. HÖRMANDER* : On the uniqueness of the Cauchy problem II. Math. Scand. 7 (1959), 177-190.

[19] *L. HÖRMANDER* : Linear partial differential operators. Springer Verlag (1963).

[20] *L. HÖRMANDER* : Non-uniqueness for the Cauchy problem. Lecture Notes in Math. Springer Verlag n°459 (1975), 36-72.

[21] *L. HÖRMANDER* : The Weyl calculus of pseudo-differential operators. Comm. Pure and Appl. Math. 32 (3) (1979), 359-443.

[22] *J. HOUNIE-M.E. MORAES MELO* : Uniqueness of the Cauchy problem for a class of differential operators with double characteristics. To appear.

[23] *V.M. ISAKOV* : On the uniqueness of the solution of the Cauchy problem. Soviet. Math. Dokl. Vol. 22 (1980) n°3, 639-642.

[24] *H.KUMANU-GO* : On the uniqueness of the Cauchy problem and unique continuation theorems for elliptic equations. Osaka Math. Journ. 14 (1962), 181-212.

[25] *H.KUMANU-GO* : Pseudo-differential operators and the uniqueness of the Cauchy problem. Comm. on Pure and Applied Math. 22 (1969), 73-129.

[26] *E.M. LANDIS* : An example of non-uniqueness of solutions of Cauchy's problem for a system of the form etc... Math. Sbornik N.S 27 (69)(1950),319-323.

[27] *R. LASCAR-C. ZUILY* : Unicité et non unicité du problème de Cauchy pour une classe d'opérateurs différentiels à caractéristiques doubles. Duke Math. Journ. (49), (1982), 137-162.

[28] *J. LERAY* : Equations hyperboliques non strictes : contr'exemples du type de Giorgi aux théorèmes d'existence et d'unicité. Math. Annalen 162 (1966), 228-236.

[29] *N. LERNER*:Unicité du problème de Cauchy pour des opérateurs elliptiques. Prépublication Université Paris Sud (Orsay)(1982).

[30] *W. MATSUMOTO* : Uniqueness of the Cauchy problem for partial differential equations with multiple characteristic roots. Journ. Math. Kyoto Univ. (1975), 479-5252.

[31] *A. MENIKOFF* : Carleman estimates for partial differential operators with real coefficients. Arch. Rat. Mech. Analysis 54 (1974), 118-133.

[32] *S. MIZOHATA* : Unicité du prolongement des solutions des équations elliptiques du quatrième ordre. Proc. Japan Acad. 34 (1958) 687-692.

[33] *S. NAKANE* : Uniqueness and non-uniqueness in the C.p. for a class of operators of degenerate type. Proc. Japan Acad. 58 Serie A (1982).

[34] *L. NIRENBERG* : Uniqueness in Cauchy problem for

differential equations with constant leading coefficients. Comm. Pure and Appl. Math. 10 (1957) 89-105.

[35] *L. NIRENBERG* : Lectures on linear partial differential equations. Conf. Board on Math. Science Texas Univ. n°17 (May 1972).

[36] *L. NIRENBERG* : Uniqueness in the Cauchy problem for a degenerate elliptic second order equation To appear.

[37] *H. OUNAIES* : Unicité du problème de Cauchy pour des opérateurs différentiels à caractèristiques non régulières. Thèse de 3e cycle. Université Paris-Sud (Orsay)(1982).

[38] *R. PEDERSON* : On the unique continuation theorem for certain second order and fourth order elliptic equations. Comm. Pure and Appl. Math. 11 (1958) 67-80.

[39] *R. PEDERSON* : Uniqueness in Cauchy's problem for elliptic equations with double characteristics. Arkiv for Math.6 (1966), 535-549.

[40] *A. PLIŠ* : The problem of uniqueness for the solutions of a system of partial differential equations. Bull. Acad. Pol. Sc. 2(1954), 55-57.

[41] *A. PLIŠ* : Non-uniqueness in Cauchy's problem for differential equations of elliptic type. Journ. Math. Mech. 9 (1960), 557-562.

[42] *A. PLIŠ* : A smooth linear elliptic differential equation without any solution in a sphere. Comm. Pure and Appl. Math.14 (1961), 599-617.

[43] *A. PLIŠ* : Unique continuation theorems for solutions of partial differential equations. Proc. Intern. Congr. Math. Stockholm (1962), 397-402.

[44] *A. PLIŠ* : On non-uniqueness in Cauchy problem for an elliptic second order differential equation. Bull. Acad. Pol. Sc. 11 (1963), 95-100.

[45] *A. PLIS* : Homogeneous partial differential equations possessing solutions with arbitrary small supports. Bull. Acad. Pol. Sc. 12 (1964), 205-206.

[46] *M. PROTTER* : Unique continuation for elliptic equations. Trans. Amer. Math. Soc. 95 (1960), 81-91.

[47] *G.B. ROBERTS-P. WENSTON* : Uniqueness of the Cauchy problem for weakly hyperbolic operators not satisfying Levi conditions. Journ. of Diff. Equ. 40 7-36 (1981).

[48] *X. SAINT-RAYMOND* : L'unicité pour des problèmes de Cauchy caractèristiques. Comm. on p.d.e. Vol. 7 n° 5 (1982), 559-579.

[49] *T. SHIROTA* : A remark on the unique continuation theorem for certain fourth order elliptic equations. Proc. Japan Acad. 36 (1960), 571-573.

[50] *T. SMITH* : Some remarks on a paper of Calderon on existence and uniqueness theorems for systems of p.d.e. Comm. Pure Appl. Math. 18 (1965), 415-441.

[51] *M. STRAUSS-F. TREVES* : First order linear pde's and uniqueness of the Cauchy problem. Journ. of Diff. Equations 15 (1974) 195-209.

[52] *M. SUSSMAN* : On the uniqueness in Cauchy's problem for elliptic operators with characteristics of multiplicity greater than two. Bull. Amer. Math. Soc. 81 (1975).

[53] *F. TREVES* : A link between solvability of pseudo-differential equations and uniqueness of the Cauchy problem. Amer. Journ. of Math. 94 (1972), 267-288

[54] *F. TREVES* : Approximation and representation of functions and distributions annihilated by a system of complex vector fields. Centre de Math. Ecole Polytechnique 91128 Palaiseau (France) (Mai 1981).

[55] *K. WATANABE* : On the uniqueness of the Cauchy problem for certain elliptic equation with triple characteristics. Tohoku Math. Journ. 23(1971), 473-490.

[56] *K. WATANABE* : A unique continuation theorem for an elliptic operator of two independant variables with non-smooth characteristics. Osaka Journ. of Math. 10 (1973), 243-246.

[57] *K. WATANABE* : Sur l'unicité du problème de Cauchy pour les opérateurs différentiels à caracteristiques de multiplicité constante. Tohoku Math. Joum. 31 (1979), 151-164.

[58] *K. WATANABE* : L'unicité du prolongement des solutions des équations elliptiques dégénérées. Tohoku Math. Journ. 34 (1982), 239-249.

[59] *K. WATANABE-C. ZUILY* : On the uniqueness of the Cauchy problem for elliptic differential operators with smooth characteristics of variable multiplicity. Comm. on p.d.e 2 (8) (1977), 831-855.

[60] *H. WHITNEY* : Analytic extensions of differentiable functions defined in closed set. Trans. Amer. Math. Soc. 36(1934), 63-89.

[61] *M. ZEMAN* : Uniqueness of solutions of the Cauchy problem for linear partial differential equations with characteristics of constant multiplicity. Journ. Diff. Equations 24(1977) 178-196.

[62] *M. ZEMAN* : Uniqueness of solutions of the Cauchy problem for linear partial differential equations with characteristics of variable multiplicity. Journ. Diff. Equations 27 (1978), 1-18.

[63] *M. ZEMAN* : On the uniqueness of Cauchy problem for partial differential operators with multiple characteristics. Annali Scuola Norm. Sup. Pisa IV Vol VII n°2 (1980), 257-285.

[64] *C. ZUILY* : Unicité du problème de Cauchy pour une classe d'opérateurs différentiels. Comm. on p.d.e Vol 6 n°2, (1981), 153-196.

[65] *C. ZUILY* : Second order elliptic equations and the uniqueness of the Cauchy problem. To appear in Bull. Soc. Math. Brasiliera.

Progress in Mathematics

Edited by J. Coates and S. Helgason

Progress in Physics

Edited by A. Jaffe and D. Ruelle

- A collection of research-oriented monographs, reports, notes arising from lectures or seminars
- Quickly published concurrent with research
- Easily accessible through international distribution facilities
- Reasonably priced
- Reporting research developments combining original results with an expository treatment of the particular subject area
- A contribution to the international scientific community: for colleagues and for graduate students who are seeking current information and directions in their graduate and post-graduate work.

Manuscripts

Manuscripts should be no less than 100 and preferably no more than 500 pages in length.

They are reproduced by a photographic process and therefore must be typed with extreme care. Symbols not on the typewriter should be inserted by hand in indelible black ink. Corrections to the typescript should be made by pasting in the new text or painting out errors with white correction fluid.

The typescript is reduced slightly (75%) in size during reproduction; best results will not be obtained unless the text on any one page is kept within the overall limit of 6x9½ in (16x24 cm). On request, the publisher will supply special paper with the typing area outlined.

Manuscripts should be sent to the editors or directly to:
Birkhäuser Boston, Inc., P.O. Box 2007, Cambridge, Massachusetts 02139

PROGRESS IN MATHEMATICS

Already published

PM 1 Quadratic Forms in Infinite-Dimensional Vector Spaces
Herbert Gross
ISBN 3-7643-1111-8, 432 pages, paperback

PM 2 Singularités des systèmes différentiels de Gauss-Manin
Frédéric Pham
ISBN 3-7643-3002-3, 346 pages, paperback

PM 3 Vector Bundles on Complex Projective Spaces
C. Okonek, M. Schneider, H. Spindler
ISBN 3-7643-3000-7, 396 pages, paperback

PM 4 Complex Approximation, Proceedings, Quebec, Canada, July 3-8, 1978
Edited by Bernard Aupetit
ISBN 3-7643-3004-X, 128 pages, paperback

PM 5 The Radon Transform
Sigurdur Helgason
ISBN 3-7643-3006-6, 202 pages, paperback

PM 6 The Weil Representation, Maslov Index and Theta Series
Gérard Lion, Michèle Vergne
ISBN 3-7643-3007-4, 348 pages, paperback

PM 7 Vector Bundles and Differential Equations
Proceedings, Nice, France, June 12-17, 1979
Edited by André Hirschowitz
ISBN 3-7643-3022-8, 256 pages, paperback

PM 8 Dynamical Systems, C.I.M.E. Lectures, Bressanone, Italy, June 1978
John Guckenheimer, Jürgen Moser, Sheldon E. Newhouse
ISBN 3-7643-3024-4, 298 pages, paperback

PM 9 Linear Algebraic Groups
T. A. Springer
ISBN 3-7643-3029-5, 314 pages, hardcover

PM10 Ergodic Theory and Dynamical Systems I
A. Katok
ISBN 3-7643-3036-8, 346 pages, hardcover

PM11 18th Scandinavian Congress of Mathematicians, Aarhus, Denmark, 1980
Edited by Erik Balslev
ISBN 3-7643-3040-6, 526 pages, hardcover

PM12 Séminaire de Théorie des Nombres, Paris 1979-80
Edited by Marie-José Bertin
ISBN 3-7643-3035-X, 404 pages, hardcover

PM13 Topics in Harmonic Analysis on Homogeneous Spaces
Sigurdur Helgason
ISBN 3-7643-3051-1, 152 pages, hardcover

PM14 Manifolds and Lie Groups, Papers in Honor of Yozô Matsushima
Edited by J. Hano, A. Marimoto, S. Murakami, K. Okamoto, and H. Ozeki
ISBN 3-7643-3053-8, 476 pages, hardcover

PM15 Representations of Real Reductive Lie Groups
David A. Vogan, Jr.
ISBN 3-7643-3037-6, 776 pages, hardcover

PM16 Rational Homotopy Theory and Differential Forms
Phillip A. Griffiths, John W. Morgan
ISBN 3-7643-3041-4, 258 pages, hardcover

PM17 Triangular Products of Group Representations and their Applications
S.M. Vovsi
ISBN 3-7643-3062-7, 142 pages, hardcover

PM18 Géométrie Analytique Rigide et Applications
Jean Fresnel, Marius van der Put
ISBN 3-7643-3069-4, 232 pages, hardcover

PM19 Periods of Hilbert Modular Surfaces
Takayuki Oda
ISBN 3-7643-3084-8, 144 pages, hardcover

PM20 Arithmetic on Modular Curves
Glenn Stevens
ISBN 3-7643-3088-0, 236 pages, hardcover

PM21 Ergodic Theory and Dynamical Systems II
A. Katok, editor
ISBN 3-7643-3096-1, 226 pages, hardcover

PM22 Séminaire de Théorie des Nombres, Paris 1980-81
Marie-José Bertin, editor
ISBN 3-7643-3066-X, 374 pages, hardcover

PM23 Adeles and Algebraic Groups
A. Weil
ISBN 3-7643-3092-9, 138 pages, hardcover

PM24 Enumerative Geometry and Classical Algebraic Geometry
Patrick Le Barz, Yves Hervier, editors
ISBN 3-7643-3106-2, 260 pages, hardcover

PM25 Exterior Differential Systems and the Calculus of Variations
Phillip A. Griffiths
ISBN 3-7643-3103-8, 349 pages, hardcover

PM26 Number Theory Related to Fermat's Last Theorem
Neal Koblitz, editor
ISBN 3-7643-3104-6, 376 pages, hardcover

PM27 Differential Geometric Control Theory
Roger W. Brockett, Richard S. Millman, Hector J. Sussmann, editors
ISBN 3-7643-3091-0, 349 pages, hardcover

PM28 Tata Lectures on Theta I
David Mumford
ISBN 3-7643-3109-7, 254 pages, hardcover

PM29 Birational Geometry of Degenerations
Robert Friedman and David R. Morrison, editors
ISBN 3-7643-3111-9, 410 pages, hardcover

PM30 CR Submanifolds of Kaehlerian and Sasakian Manifolds
Kentaro Yano, Masahiro Kon
ISBN 3-7643-3119-4, 223 pages, hardcover

PM31 Approximations Diophantiennes et Nombres Transcendants
D. Bertrand and M. Waldschmidt, editors
ISBN 3-7643-3120-8, 349 pages, hardcover

PM32 Differential Geometry
Robert Brooks, Alfred Gray, Bruce L. Reinhart, editors
ISBN 3-7643-3134-8, 267 pages, hardcover

PROGRESS IN PHYSICS

Already published

PPh1 Iterated Maps on the Interval as Dynamical Systems
Pierre Collet and Jean-Pierre Eckmann
ISBN 3-7643-3026-0, 256 pages, hardcover

PPh2 Vortices and Monopoles, Structure of Static Gauge Theories
Arthur Jaffe and Clifford Taubes
ISBN 3-7643-3025-2, 294 pages, hardcover

PPh3 Mathematics and Physics
Yu. I. Manin
ISBN 3-7643-3027-9, 112 pages, hardcover

PPh4 Lectures on Lepton Nucleon Scattering and Quantum Chromodynamics
W.B. Atwood, J.D. Bjorken, S.J. Brodsky, and R. Stroynowski
ISBN 3-7643-3079-1, 574 pages, hardcover

PPh5 Gauge Theories: Fundamental Interactions and Rigorous Results
P. Dita, V. Georgescu, R. Purice, editors
ISBN 3-7643-3095-3, 406 pages, hardcover

PPh6 Third Workshop on Grand Unification
University of North Carolina, Chapel Hill, April 15-17, 1982
P.H. Frampton, S.L. Glashow, and H. van Dam, editors
ISBN 3-7643-3105-4, 382 pages, hardcover